introduction to
professional engineering
in canada

Gordon C. Andrews, P.Eng. | Professor Emeritus, University of Waterloo

J. Dwight Aplevich, P.Eng. | Professor, University of Waterloo

Roydon A. Fraser, P.Eng. | Associate Professor, University of Waterloo

Herbert C. Ratz, P.Eng. | Professor Emeritus, University of Waterloo

Toronto

National Library of Canada Cataloguing in Publication Data

Main entry under title:

   Introduction to professional engineering in Canada / Gordon C. Andrews ... [et al.].

Includes index.
ISBN 0-13-139745-1

1. Engineering. I. Andrews, G. C. (Gordon Clifford), 1937-

TA157.I57 2002   620   C2002-903128-1

Copyright © 2003 Pearson Education Canada Inc., Toronto, Ontario

All Rights Reserved. This publication is protected by copyright, and permission should be obtained from the publisher prior to any prohibited reproduction, storage in a retrieval system, or transmission in any form or by any means, electronic, mechanical, photocopying, recording, or likewise. For information regarding permission, write to the Permissions Department.

0-13-139745-1

Vice President, Editorial Director: Michael J. Young
Acquisitions Editor: Leslie Carson
Marketing Manager: Steve McGill
Developmental Editor: Toni Chahley
Production Editor: Mary Ann McCutcheon
Copy Editor: Betty Robinson
Production Coordinator: Patricia Ciardullo
Cover Design: Amy Harnden
Cover Image: Photodisc
Art Director: Julia Hall

2 3 4 5   07 06 05 04 03

Printed and bound in Canada.

The excerpts from Regulation 941 under the Professional Engineers Act in Section 1.1 of Chapter 3 and Section 3.1 of Chapter 3 are copyright © Queen's Printer for Ontario, 1990. This is an unofficial version of Government of Ontario legal materials.

Figure 1.3 is copyright © 2002 the authors and their licensors. All rights reserved.

The terms 386 processor, 486 DX processor, and Pentium in Figure 9.5 are trademarked terms held by the Intel Corporation.

Acknowledgement for Figure 17.3 is given to the Air-Conditioning and Refrigeration Institute and the Association of Professional Engineers and Geoscientists of New Brunswick for the u of their registered trademarks.

Preface

This book is intended to explain the elements of what every beginning engineering student should know about the engineering profession in Canada, emphasizing basic skills and knowledge that are well known to practising engineers, and particularly useful to students. The book has evolved through several versions and has served engineering students for more than a decade. This edition includes a major revision of almost all chapters to reflect current trends in engineering practice, teaching, and computer technology. The four parts of the book are organized as follows:

Part I—The engineering profession (Chapters 1 to 5): In Chapters 1, 2, and 3, students see that "engineering" is not just a set of university courses but is an organized profession, with strict requirements for admission, a code of ethics, and professional regulations. Chapter 4 explains how engineering societies are an aid to lifelong learning, and Chapter 5 gives advice for studying and writing examinations to help students in their immediate task of acquiring an engineering education.

Part II—Engineering communications (Chapters 6 to 9): In emphasizing the value of communicating effectively in print, Chapter 6 reviews document types, emphasizing technical documents that are not typically discussed elsewhere, and Chapter 7 reviews basic writing techniques. Chapter 8 contains a thorough discussion of the purpose and structure of engineering reports and the mechanics of writing them. Many students need this knowledge immediately because they may be required to write formal reports in their courses or during their first work-term or internship. Our philosophy has also been to attempt to show how to write a report that will stand up to challenge, such as might happen in a court of law. Chapter 9 describes elementary principles for the use and design of graphics and the essential rules for including graphics in engineering documents.

Part III—Engineering measurements (Chapters 10 to 14): This part explains techniques for gathering, manipulating, and presenting engineering measurements, upon which many key decisions are based. Chapter 10 contains a careful treatment of unit systems, with emphasis on SI units and rules for including them in written documents. Systematic and random measurement errors are treated in Chapter 11, together with correct use of significant digits. Chapter 12 introduces methods for estimating propagated error and provides a motivation for the elementary statistics and probability in Chapters 13 and 14.

Part IV—Engineering practice (Chapters 15 to 20): This part introduces several distinctive topics that make engineering a profession of practice, rather than of purely academic study. Chapter 15 is a basic introduction to the design process, emphasizing the iterative nature of design. Chapter 16 is a brief introduction to the importance of creativity and decision-making in design. The rapidly changing domain of intellectual property is described in Chapter 17 in order to describe how the results of the creative process can be protected and exploited. Chapter 18 briefly describes project planning and illustrates several common planning techniques using simple examples. Chapter 19 is a basic introduction to safety in engineering design and practice and lists guidelines for eliminating workplace and other hazards. Chapter 20 introduces the complex problem of calculating and managing risks associated with large systems.

How to use this book: At the University of Waterloo, we use this book in several first-year engineering courses, supplemented with videos, exercises, and assignments. Students normally produce a formal engineering report on a basic design subject or a topic that requires elementary analysis. Knowledge of the concepts is tested by several short examinations during the term. The courses also include an introduction to software tools applicable to the engineering discipline. Portions of the material—particularly the report-writing and design chapters—are also used in higher-level courses.

The four parts of the book, and most of the chapters, are designed to stand on their own, so that the order in which the topics are presented can be varied considerably. We advise introducing the professional matters in Part I early, together with an introduction to Part II if student assignments include writing engineering reports. The chapters in Part III on measurements and errors can then be covered, ending the course with chapters on design, intellectual property, or safety from Part IV. However, a different order may suit other courses.

This book is intended for beginning engineering students. One of the authors (Gordon Andrews) has co-authored a more advanced textbook, *Canadian Professional Engineering Practice and Ethics,* which is intended for senior courses, practising engineers, and graduates preparing to write provincial professional engineering practice exams (see reference [6] in Chapter 1).

Acknowledgements: The authors would like to thank all those who assisted, directly or indirectly, in the writing of this book. We gratefully acknowledge the students, teaching assistants, and many colleagues who provided suggestions, comments, and criticism of previous versions of the book. Surin Kalra, Bob Pearce, and the late Alan Hale assisted immensely and are gratefully remembered. Comments and suggestions from Roy Pick, Rob Gorbet, Carolyn MacGregor, Ewart Brundrett, John Shortreed, Carl Thompson, Neil Thomson, Barry Wills, June

Lowe, Don Fraser, and others were very much appreciated. The Director of First-Year Studies at the University of Waterloo created a positive environment for the development of this book, and we would like to thank Jim Ford, Gord Stubley, and Bill Lennox who held this position in recent years.

We are indebted to people who helped us to get permission to use copyright material and to the copyright holders. In particular, we would like to thank Connie Mucklestone and Johnny Zuccone of PEO, Melissa Mertz of APEGNB, Lynne Vanin of MD Robotics, Paul Deslauriers of Waterloo FSAE, Nick Farinaccio of Chipworks, Richard Van Vleck of American Artifacts, and Mark Menzer of ARI.

We thank Connie Mucklestone of PEO for her thorough reading and detailed comments on Part I of the book.

We owe thanks to the people at Pearson Education Canada who helped us to meet a tight publishing schedule: editors Kelly Torrance, Leslie Carson, Toni Chahley, Mary Ann McCutcheon, and others. Two people deserve special thanks: Paul McInnis for his perspicacious suggestions, and Betty Robinson for her very thorough reading of the text, diagrams, and mathematics.

Every effort has been made to avoid errors and to credit sources. The authors would be grateful for advice concerning any errors or omissions. In particular, we would like to know, from students, what you particularly liked or disliked in this book and, from instructors, what changes you would like to see in later editions. Please contact us via the web site <http://ece.uwaterloo.ca/~aplevich/introeng>.

Gordon C. Andrews
J. Dwight Aplevich
Roydon A. Fraser
Herbert C. Ratz

June 2002

Contents

Preface, *iii*
List of figures, *xiv*
List of tables, *xvii*

Part I The engineering profession — 1

Chapter 1 An introduction to engineering — 3

1. What is an engineer? *3*
2. The role of the engineer, *5*
3. Engineering disciplines, *6*
4. Becoming a competent engineer, *10*
5. Challenges for engineering, *12*
6. Further study, *13*
7. References, *14*

Chapter 2 The licensed professional engineer — 15

1. Engineering is a profession, *16*
2. Regulation of the engineering profession, *16*
   - 2.1 The laws regulating engineering, *17*
   - 2.2 The legal definition of engineering, *19*
3. Admission to the engineering profession, *20*
   - 3.1 Academic requirements, *21*
   - 3.2 Experience requirements, *22*
   - 3.3 The professional practice examination, *23*
   - 3.4 Certificates of authorization, *23*
4. Provincial engineering organizations, *24*
5. Further study, *24*
6. References, *25*

Chapter 3 Professional engineering ethics — 27

1. Introduction to professional ethics, *27*
   1.1 PEO Code of Ethics, *28*
2. Ethics in the workplace, *30*
3. Professional misconduct and discipline, *32*
   3.1 Definition of professional misconduct, *33*
4. The professional use of computer programs, *34*
5. Proper use of the engineer's seal, *36*
6. The iron ring and the engineering oath, *37*
7. Further study, *38*
8. References, *40*

Chapter 4 Engineering societies — 41

1. The purpose of engineering societies, *41*
2. The history of engineering societies, *42*
3. The importance of engineering societies, *43*
4. Choosing your engineering society, *45*
   4.1 Canadian engineering societies, *46*
   4.2 American and international engineering societies, *47*
5. Further study, *48*
6. References, *48*

Chapter 5 Advice on studying and exams — 49

1. The good and bad news about university studies, *49*
2. How much study time is required? *50*
3. Preparing for the start of lectures, *50*
4. A checklist of good study skills, *51*
5. Developing a note-taking strategy, *52*
6. Hints for assigned work, *53*
7. Preparing for examinations, *53*
8. Writing examinations, *54*
9. When things go wrong, *55*
10. Further study, *55*
11. References, *56*

Part II Engineering communications — 57

Chapter 6 Technical documents — 59

1. Types of technical documents, *59*
   1.1 Letters, *60*
   1.2 Memos, *60*
   1.3 Email, *62*

Contents ix

    1.4   Specification documents, *63*
    1.5   Bids and proposals, *64*
    1.6   Reports, *64*
2   Further study, *68*
3   References, *68*

Chapter 7   Technical writing basics   69

1   The importance of clarity, *69*
2   Hints for improving your writing style, *70*
3   Punctuation: A basic summary, *73*
4   The parts of speech: A basic summary, *75*
5   Avoid these writing errors, *75*
6   The Greek alphabet in technical writing, *77*
7   Further study, *77*
8   References, *80*

Chapter 8   Formal technical reports   81

1   Components of a formal report, *82*
    1.1   The front matter, *82*
    1.2   The report body, *87*
    1.3   The back matter, *90*
2   Steps in writing a technical report, *93*
3   A checklist for engineering reports, *95*
4   Further study, *97*
5   References, *98*

Chapter 9   Report graphics   99

1   Graphics in engineering documents, *100*
2   Standard formats for graphs, *100*
    2.1   Bar charts and others, *102*
    2.2   Straight-line graphs, *103*
    2.3   Logarithmic scales, *104*
3   Engineering calculations, *107*
4   Sketches, *109*
5   Further study, *110*
6   References, *112*

Part III   Engineering measurements   113

Chapter 10   Measurements and units   115

1   Measurements, *115*
2   Unit systems for engineering, *116*

3 Writing quantities with units, *118*
4 Basic and common units, *119*
5 Unit algebra, *121*
6 Further study, *122*
7 References, *124*

Chapter 11 Measurement error — 125

1 Measurements, uncertainty, and calibration, *125*
2 Systematic and random errors, *126*
   2.1 Systematic errors, *127*
   2.2 Random errors, *127*
3 Precision, accuracy, and bias, *128*
4 Estimating measurement error, *129*
5 How to write inexact quantities, *129*
   5.1 Explicit uncertainty notation, *129*
   5.2 Implicit uncertainty notation, *130*
6 Significant digits, *130*
   6.1 Rounding numbers, *131*
   6.2 The effect of algebraic operations, *132*
7 Further study, *134*
8 References, *134*

Chapter 12 Error in derived quantities — 135

1 Method 1: Exact range of a calculated result, *135*
2 Method 2: Linear estimate of the error range, *137*
   2.1 Sensitivities, *139*
   2.2 Relative sensitivities, *140*
   2.3 Approximate error range, *141*
   2.4 Application of Method 2 to algebraic functions, *142*
3 Method 3: Estimated uncertainty, *144*
   3.1 Derivation of the estimated value, *146*
4 Further study, *147*
5 References, *150*

Chapter 13 Statistics — 151

1 Basic definitions, *151*
   1.1 Engineering applications of statistics, *152*
   1.2 Descriptive and inferential statistics, *153*
2 Measures of central value, *155*
   2.1 Mean, median, and mode of skewed distributions, *158*
3 Measures of spread, *159*
4 Measures of relative standing, *161*
   4.1 Measures of relative standing spread, *161*

Contents xi

    5   Presentation of measured data, *161*
        5.1   Histogram interval selection, *163*
    6   Further study, *163*
    7   References, *164*

Chapter 14   Gaussian law of errors   165

    1   The Gaussian law of errors, *165*
        1.1   Conditions for the Gaussian law of errors, *168*
    2   The Gaussian error distribution, *169*
        2.1   Properties of the sample mean, *172*
    3   Validating the Gaussian law of errors, *173*
        3.1   Histogram of observed data, *173*
    4   The best straight line, *175*
        4.1   Correlation coefficient, *176*
    5   Rejection of an outlying point, *177*
        5.1   Standard deviation test, *178*
        5.2   Application of outlier tests to skewed distributions, *179*
    6   Further study, *179*
    7   References, *180*

Part IV   Engineering practice   181

Chapter 15   The engineering design process   183

    1   Definition and importance of engineering design, *183*
    2   The design process, *184*
        2.1   Recognition of the need, *185*
        2.2   Definition of the design problem, *185*
        2.3   Defining the design criteria and constraints, *186*
        2.4   The design loop, *187*
        2.5   Optimization, *188*
        2.6   Evaluation, *188*
        2.7   Communication of the design, *188*
    3   Two design examples, *188*
    4   Design team organization, *191*
    5   Further study, *192*
    6   References, *194*

Chapter 16   Creativity and decision-making   195

    1   Synthesis: The creative process, *195*
    2   Characteristics of creative people, *196*
    3   Stimulating engineering creativity, *197*
        3.1   Brainstorming, *197*

        3.2   Brainwriting, *198*
        3.3   Overcoming creative blocks, *198*
4   Analysis, *199*
5   Decision-making, *199*
        5.1   A basic decision-making method, *200*
        5.2   Computational decision-making, *200*
6   References, *202*

Chapter 17   Intellectual property   203

1   Introduction, *203*
        1.1   Proprietary intellectual property, *204*
        1.2   The public domain, *206*
2   The importance of intellectual property, *206*
        2.1   Rights of employers and employees, *207*
3   Patents, *208*
        3.1   Patent developments, *209*
        3.2   The patent application process, *210*
4   Copyright, *212*
        4.1   Copyright registration, *212*
        4.2   Fair dealing, *213*
        4.3   Copyright and computer programs, *213*
5   Industrial designs, *214*
6   Trademarks, *214*
7   Integrated circuit topographies, *216*
8   Trade secrets, *217*
9   Further study, *217*
10   References, *219*

Chapter 18   Project planning and scheduling   221

1   Gantt charts and the critical-path method, *221*
2   Planning with CPM, *222*
3   Scheduling with CPM, *225*
4   Refinement of CPM, *228*
5   Summary of steps in CPM, *230*
6   References, *232*

Chapter 19   Safety in engineering design   233

1   Responsibility of the design engineer, *233*
2   Some general guidelines, *234*
3   Principles of hazard recognition and control, *236*
4   Eliminating workplace hazards, *238*
5   Cost-benefit justification of safety issues, *239*
6   Codes and standards, *240*

Contents **xiii**

      6.1   Finding and using safety codes and standards, *241*
7   Further study, *243*
8   References, *244*

Chapter 20 Safety, risk, and the engineer   245

1   Evaluating risk in design, *245*
2   Risk management, *246*
3   Analytical methods, *247*
      3.1   Checklists, *247*
      3.2   Hazard and operability (HAZOP) studies, *248*
      3.3   Failure modes and effects analysis (FMEA), *249*
      3.4   Fault-tree analysis, *251*
4   Safety in large systems, *253*
5   System risk, *254*
6   Expressing the costs of a hazard, *255*
7   Further study, *256*
8   References, *258*

Index, *259*

List of figures

I.1   The Canadarm, *1*

1.1   The iron ring, *3*
1.2   Names of accredited Canadian engineering programs, *8*
1.3   Engineering faces many challenges, *12*

2.1   The collapse of the Québec bridge, *15*
2.2   A typical engineering licence, *18*
2.3   The steps to take to become an engineer via an accredited program, *20*

3.1   A typical professional engineer's seal, *37*

4.1   A Formula SAE race, *41*
4.2   The relationship of the licensed engineer to engineering organizations, *44*

5.1   A large university lecture hall, *49*

II.1 A flowchart of a manufacturing process, *57*

6.1   An antique typewriter, *59*
6.2   The format of a typical business letter, *61*
6.3   A typical memo, *62*

7.1   Part of an entry from a thesaurus, *69*
7.2   A general-purpose latch, *70*

8.1   A well-produced technical report makes a subtle statement, *81*
8.2   A typical report cover defined by a company, *84*

9.1   Napoleon's campaign in Russia, *99*
9.2   The standard format for graphs, *101*
9.3   A bar chart and a pie chart comparing six quantities, *102*

9.4 A straight-line function, *103*
9.5 Moore's law, *104*
9.6 A graph with linear scales and a log-linear plot of a decaying exponential, *105*
9.7 A power function produces a straight line on a log-log plot, *106*
9.8 A standard format for engineering calculations, *108*
9.9 A sketch conveys essential details informally, *110*

III.1 A railroad transit from the Bausch & Lomb 1908 Catalog of Engineering Instruments, *113*

11.1 A modern vernier caliper, *125*
11.2 True value, measured estimate, and interval of uncertainty, *126*
11.3 Precision, bias, and accuracy, *128*

12.1 Computing the extremes of a function over an interval, *137*
12.2 The change in $f(x)$ is approximately the slope of f at x_0 multiplied by the deviation Δx, *138*
12.3 Graphs of Δx_k, Δy_k, the squared quantities, and the products $\Delta x_k \Delta y_y$, *146*

13.1 Language test mark distribution for all tests, *155*
13.2 The distribution of salaries in a small company, *157*
13.3 A negatively skewed distribution, *159*
13.4 Metal-fatigue histogram from Example 3, *163*

14.1 The Gaussian (or normal) distribution with mean μ and standard deviation σ, *166*
14.2 The occurrences of heads in n coin flips, *167*
14.3 The Gaussian distribution with intervals of percentage area, *169*
14.4 The cumulative Gaussian error distribution function, *171*
14.5 Histogram of metal-fatigue test data from Example 3, *174*
14.6 A nonlinear transformation changes the relative size of error ranges, *176*

IV.1 A construction site, *181*

15.1 A computer circuit board, *183*
15.2 The design process, showing the many possibilities for iteration, *185*
15.3 An automobile assembly line, *190*

17.1 Microchip humour, *203*
17.2 A self-congratulatory apparatus, *209*
17.3 Certification marks from the CIPO database, *215*

xvi List of figures

18.1 A basic Gantt chart for a design project, *222*
18.2 Activities and events in an arrow diagram, *223*
18.3 Arrow diagram for a one-operator tire purchase, *224*
18.4 Arrow diagram for a large-store tire purchase, *224*
18.5 Arrow diagram for the improved procedure, *226*
18.6 Format for an event circle, *226*
18.7 Arrow diagram for one operator, from Figure 18.3, *229*
18.8 Arrow diagram for the large store, from Figure 18.4, *229*
18.9 The improved procedure, from Figure 18.5, *231*
18.10 Gantt chart for the improved procedure of Figure 18.9, *231*

19.1 Examples of standard hazard signs, *236*

20.1 Partial fault tree for the hair dryer in Example 2, *252*
20.2 Illustrating lines of constant risk, *255*
20.3 Compressed-air supply unit, *257*
20.4 Part of the fault tree for Problem 1, *257*

List of tables

6.1   Components of a typical laboratory report, *67*

7.1   Tenses for the verb *to see* in both active and passive voice, *72*
7.2   The Greek alphabet, showing English equivalents, *77*

8.1   Components of a formal report, *83*

10.1   Base and derived SI units, *117*
10.2   Comparison of unit systems, *118*
10.3   Magnitude prefixes in the SI unit system by symbol, name, and numerical value, *120*

13.1   Examples of description and inference, *154*
13.2   Mark occurrences and relative frequencies for past language tests, *154*
13.3   Observations from 662 paper clips fatigue-tested to failure, *162*

14.1   Standard Gaussian distribution function interval probabilities, *170*

16.1   Payoff computation for the student-travel problem, *202*

17.1   Type, criteria for protection, and term of intellectual property, *205*

18.1   Activities for the sale and installation of winter tires, *225*

19.1   Examples of hazard control methods., *235*
19.2   Hazard checklist for machine design, *240*
19.3   Hazard checklist for workplace layout, *241*

Part I
The engineering profession

Fig. I.1. Design of the Canadarm and Canadarm 2 required expertise in several engineering branches: materials, structures, robotics, electronics, control systems, computers, and software. The Canadarm is listed among the five most significant Canadian engineering achievements of the 20th century. This is an artist's rendering of the mobile servicing system, which includes the Canadarm 2 and other elements. (Courtesy of MD Robotics)

By choosing engineering, you have taken the first step to a challenging and rewarding profession. In future years you will share the great sense of personal achievement that is typical of engineering, as your ideas move from the design office or computer lab to the production line or construction site. This textbook is intended to acquaint future engineers with many, many basic engineering concepts. However, in passing along this basic information, we hope that the excitement and creativity of engineering also show through.

Part I introduces the reader to the engineering profession.

Chapter 1—An introduction to engineering: Most people know that engineers wear iron rings, but how is engineering defined, exactly? This chapter gives you a working definition of engineering and explains the difference between engineers and other technical specialists such as research scientists, technicians and technologists.

Chapter 2—The licensed professional engineer: An engineering education is not just a set of related university courses; it is the entry point to a legally recognized profession with strict requirements for admission, a code of ethics, and professional regulations. This chapter tells you how the engineering profession is organized and how you get into it.

Chapter 3—Engineering ethics: The public expects all professional people to be honest, reliable and ethical. How does the public expectation apply to engineering? This chapter describes the Code of Ethics, how it applies to typical engineering practice, the proper use of the engineer's seal and the significance of the iron ring.

Chapter 4—Engineering societies: Engineering societies help you by publishing technical papers, organizing conferences and engineering contests, and by presenting short courses. Such assistance is very useful to engineering undergraduates, but it is even more important after you graduate. This chapter explains the role of engineering societies and describes several societies that may be of interest to you.

Chapter 5—Advice on studying and exams: You are investing a lot of time and money in your engineering education. Would you like to protect this investment? If you can master the fundamental advice in this chapter, you should be able to guarantee academic success and still have enough free time to enjoy the many other interesting aspects of university life.

Chapter 1

An introduction to engineering

The iron ring shown in Figure 1.1 is worn with pride by graduates of Canadian engineering degree programs, but the wearing of an iron ring does not make a person an engineer. As a first glimpse of what might lie ahead in your engineering career, this chapter gives a basic introduction to engineering; in it, you will learn:

- a working definition of the word *engineer*,
- the role of engineers compared to other technical specialists,
- elements of some of the common engineering disciplines,
- the talents and skills needed to become an engineer, and
- some of the engineering challenges faced by society.

The Further study section at the end of the chapter poses some questions that may help you to analyse and understand your personal career choices and challenges.

Fig. 1.1. The iron ring worn by graduates of Canadian engineering degree programs.

1 What is an engineer?

The term *engineer* comes to English from the Latin word *ingenium,* meaning talent, genius, cleverness, or native ability. Its first use was to designate persons who had an ability to invent and operate weapons of war. Later, the word came to be associated with the design and construction of works, such as ships, roads, canals and bridges, and the people skilled in these fields were non-military, or *civil* engineers. The meaning of the term *engineering* depends, to some degree, on the country. In England, persons with artisanal skill or knowledge have been called engineers since the time of the Industrial Revolution. American usage places more emphasis on formal training, as a result of a recognized need for trained engineers during the wars of the 1700s and 1800s, and the modelling of early American engineering programs on French engineering schools [1].

In Canada, the title Professional Engineer is restricted by law to mean those persons who have demonstrated their competence and have been licensed by a provincial licensing body, which is referred to in this book as a *provincial Association*. Exceptions are permitted for stationary and sometimes military engineers, who are subject to other regulations. The legal definition of engineering and the licensing of engineers are discussed in Chapter 2. To be more precise, the terms *engineer, engineering, professional engineer, P.Eng., consulting engineer* and their French equivalents are official marks held by the Canadian Council of Professional Engineers on behalf of the provincial Associations of Professional Engineers and the Ordre des ingénieurs du Québec.

Although many complex engineering works were built by the important civilizations of antiquity, tools and techniques evolved rapidly during the 18th and 19th centuries, as a result of the intense development of machinery during the Industrial Revolution. Advances in mathematics and science permitted the prediction of strength, motion, flow, power, and other quantities with increasing accuracy. The computer revolution began in the 1950s, and the first personal computer appeared in 1975. The volume of engineering knowledge is now said to be doubling in less than every 20 years [2], and the role of engineers in modern society is rapidly evolving. Therefore, a modern definition of engineering must be broad enough to allow for change. The following is adequate in some contexts:

> *An engineer is a person who uses science, mathematics, experience, and judgement to create, operate, manage, control, or maintain devices, mechanisms, processes, structures, or complex systems, and who does this in a rational and economic way.*

The above definition is accurate, but it does not adequately express the human context of engineering as a profession. Although science, mathematics, and logic are the basis of engineering knowledge, real projects require the human skills of leadership, management, and communication. Engineers face the exciting, but occasionally unnerving responsibility of making key decisions under uncertainty of the outcome. As Petroski emphasized in *To Engineer is Human* [3], it is human nature to extend designs to their limits, where new forms of failure sometimes occur, and to learn from failures by creating new design methods.

Engineers are builders and problem solvers, and these activities give great pleasure to many people. Florman, in *The Existential Pleasures of Engineering* [4], suggests that since the time when humankind first began to use tools, the impulse to change the world around us has been part of our nature. Thus, he argues, to be human is to engineer.

Successful engineering creates and exercises interpersonal relationships, and there are many opportunities for friendship. An engineer must be a scientist and a mathematician, but must also be creative, able to communicate with and motivate others, and capable of making decisions and solving new problems.

2 The role of the engineer

Engineering is usually a team activity. Because of the great complexity of engineering projects, engineering teams often include persons with widely different abilities, interests, and education, who co-operate by contributing their particular expertise to advance the project. Although engineers are only one component of this diverse group, they contribute a vital link between theory and practical application. A typical technical team might include scientists, engineers, technologists, technicians, and skilled workers, whose activities and skills may appear at first glance to overlap and be interchangeable. The following paragraphs describe the tasks performed by different members of a typical engineering team. These are broad categorizations, and exceptions are common.

The typical **research scientist** works in a laboratory, on problems that expand the frontiers of knowledge, but which may not have practical applications for many years. A doctorate degree is usually the basic educational requirement, although a master's degree is sometimes acceptable. The research scientist typically supervises research assistants, and will usually be a member of several learned societies in his or her particular field of interest, but will not usually be a member of a self-regulating profession. The *raison d'être* of pure science is the understanding of natural phenomena, and in a project team, the main responsibility of the scientist is to provide the required scientific theories. Although many scientists apply their knowledge to practical needs, generally their profession is not subject to the same constraints or responsibilities in law as engineering.

The **engineer** typically provides the key link between theory and practical applications. The engineer must have a combination of extensive theoretical knowledge, the ability to think creatively, the knack of obtaining practical results, and the ability to lead a team toward a common goal. The bachelor's degree is the basic education requirement, although the master's or doctorate degree is useful and preferred by some employers. In Canada, a person taking responsibility for engineering work affecting public safety is required, by provincial law, to be a member or licensee of the provincial Association of engineers. Membership confers the right to use the title Professional Engineer (ing. in Québec, P.Eng. elsewhere). Not all engineers exclusively do work that is regulated by law, but work defined as engineering inherently involves professional responsibility for the results.

The **technologist** typically works under the direction of engineers in applying engineering principles and methods to complex technical problems. The basic educational requirement is graduation from a three-year technology program at a community college or equivalent, although occasionally a technologist may have a bachelor's degree, usually in science, mathematics or related subjects. The technologist often supervises the work of others and is encouraged to have qualifications recognized by a technical society. In Ontario, for example, the Ontario Association of Certified Engineering Technicians and Technologists (OACETT) confers the title of Certified Engineering Technologist (C.E.T.). This is a voluntary organization, and the title is beneficial, but not legally essential, for working as a technologist. The fundamental difference between technologists and engineers is usually the greater theoretical depth of the engineering education and the greater hands-on experience implied by the technology diploma.

The **technician** typically works under the supervision of an engineer or technologist in the practical aspects of engineering, such as making tests and maintaining equipment. The basic educational requirement is graduation from a two-year technician program at a community college or its equivalent. Associations such as OACETT may confer the title Certified Engineering Technician (C.Tech.) on qualified technicians, although the title is not essential to obtain work as a technician.

The **skilled worker** typically carries out the designs and plans of others. Such a person may have great expertise acquired through formal apprenticeship, years of experience, or both. Most trades (electrician, plumber, carpenter, welder, pattern maker, machinist, and others) have a trade organization and certification procedure.

Each of the above groups has a different task, and there are considerable differences in the skills, knowledge and performance expected of each. In particular, a much higher level of accountability is expected from the professional engineer than from other members of the engineering team. The engineer is responsible for competent performance of the work that he or she supervises. In fact, engineers may be held legally accountable not only for their own acts but also for advice given to others. Judgement and experience are often as important as mathematics, science, and technical knowledge; and liability insurance is becoming essential in the public practice of engineering. These professional aspects are discussed further in Chapter 2.

As you begin your engineering career, you should know that the team categories discussed above are not rigid barriers. Movement from one to another is always possible, although not always easy.

3 Engineering disciplines

Most people can name a few branches of engineering: civil, electrical, mechanical, and chemical engineering, perhaps. However, the number of specialized branches of engineering is much larger than is commonly known. A recent official list [5] contains the names of 224 accredited Canadian engineering degree programs. After equating equivalent French and English designations and removing duplicates, there are 69 different engineering specializations available across the country. The list changes slightly from year to year, as new programs are created or dropped, or their names are changed. Figure 1.2 shows the names occurring more than once and the number of occurrences of each name on the list. Thus, although there are 69 different specializations, 48 of them occur only once in the list, whereas electrical engineering and mechanical engineering or their French equivalents can be found in 28 universities. The figure includes the first

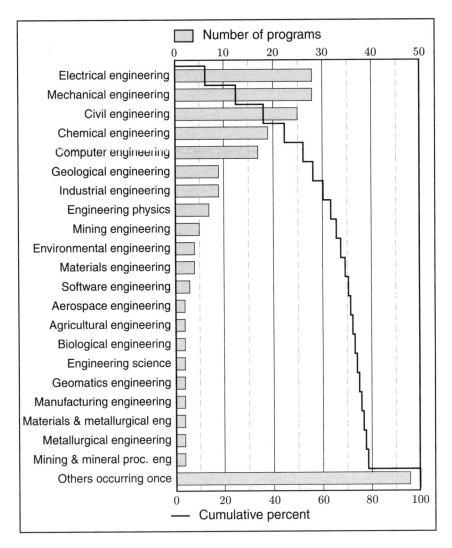

Fig. 1.2. Names of accredited Canadian engineering programs. Electrical and mechanical engineering are the two largest groups, each occurring 28 times. The cumulative percent line shows that more than 50% of the programs are in five disciplines.

three software engineering programs that were accredited in 2001 prior to release of the list. In the United States there are approximately eight times as many accredited engineering degree programs as in Canada, with twice as many program names, but the most common six names are identical to the Canadian names and have the same order as in Figure 1.2.

You must choose the engineering discipline that is right for you. The best method is to base your choice on the type of work that you want to do after graduation. Most students choose their engineering discipline before they apply for university, but if you have any doubts about your discipline, you should get further advice from your professors, guidance counsellors, or perhaps best of all, a friend who is a practising engineer. Libraries also have many references, such as encyclopaedias [2], university calendars showing the courses offered for each discipline, and textbooks [6]. The internet is also an important source of information and advice; see [7] for example.

Fundamental subjects are usually studied at the beginning of a university engineering program, including computers and information-processing systems, which are used extensively by all engineering branches. However, in higher years, students study specialized subjects that differ dramatically from branch to branch. To help you confirm the vision of engineering that you have chosen, the following paragraphs describe the work typically performed in the well-known branches of engineering.

Chemical engineers use knowledge of chemistry, physics, biology, and mathematics to design equipment and processes for the manufacture of chemicals and chemical products. The chemical process industries are skill- and capital-intensive, and require highly specialized designers to achieve competitive manufacturing. Some chemical engineers design equipment and processes for the extraction of minerals, petroleum, and other natural resources. Others design new processes for sustaining and reclaiming the natural environment.

Civil engineers design and supervise the construction of roads, highways, bridges, dams, airports, railways, harbours, buildings, water supply systems and sewage systems. Many civil engineers specialize in one phase of their discipline, such as highway, sanitary, soil, structural, transportation, or hydraulic engineering. The largest number of licensed engineers in private practice is in civil engineering.

Computer engineers use a combination of electrical engineering and computer science subjects to design, develop, and apply computer systems. They must be completely familiar with the basic aspects of electric circuits, integrated circuit design, computer hardware and architectures, software development, and numerical algorithms, and must have the ability to select the optimal combination of hardware and software components required to achieve specified system-performance criteria.

Electrical engineers design and supervise the construction of systems to generate, transmit, control and use electrical energy. Specialists in the power field typically design and develop heavy equipment, such as generators and motors, and transmission lines and distribution systems, including the complex equipment needed to control these powerful devices. Specialists in electronics and communications design and develop devices and systems for transmitting data, solid-state switching, microwave relays, computer logic circuits and computer hardware.

Environmental engineers respond to needs for improved air and water quality and efficient waste management. Environmental engineering is a rapidly growing discipline, concerned with site assessment and approval, air quality in buildings, monitoring and control devices, and a host of investigative, instrumentation, and other support activities. These environmental applications require a broad engineering knowledge, including topics from chemistry, physics, soils engineering, mechanical design, fluid mechanics, and meteorology.

Geological engineers use knowledge of the origin and behaviour of geological materials to design structures, such as foundations, roads, or tunnels built on or through these materials, and to develop exploration or extraction methods for petroleum and minerals. They may also work with civil engineers in the geotechnical design or construction of roads, airports, harbours, waste disposal systems, and other civil works. Geological engineers are also in increasing demand for ground-water and environmental impact studies and for new petroleum recovery projects such as tar-sands oil recovery.

Industrial engineers use probability, statistics, and other mathematical subjects for the design of efficient manufacturing processes. They assume responsibility for quality control, plant design, the allocation of material, financial, and human resources for efficient production. Automation, materials handling, environmental protection, robotics, human factors, and data processing are some of the specialized subjects required.

Mechanical engineers enjoy a wide scope of activity, including the design, development, manufacture, sale or maintenance of machinery, ranging from appliances to aircraft. They may be involved with engines, turbines, boilers, pressure vessels, heat exchangers, or machine tools. They may specialize in fields such as machine design, heating, ventilating and refrigeration, thermal and nuclear power generation, manufacturing, quality control or production scheduling. There are many mechanical engineers employed in related fields, such as mining, metallurgy, transportation, oil refining and chemical processing.

Software engineering is an emerging discipline that combines classical engineering project-management skills with the specialized tools and knowledge that are required to design, build, analyse, and maintain complex computer software. In addition to intimate knowledge of the program-design aspects of computer science, the software engineer requires a sufficiently broad understanding of the natural sciences, in order to work with specialists in other disciplines, and to design correct software to be embedded in machines.

Some engineering programs have a relatively broad but intense set of requirements, with the opportunity to choose specialized options or projects in later years. **Systems design engineering, engineering science, and génie unifié** are examples of such programs. Graduates may work in interdisciplinary areas requiring computer expertise and broad problem-solving and group management abilities. Some engineering disciplines are industry-related, for example **aerospace, agricultural, biosystems, building, forest, mining,** and **petroleum engineering.** Finally, some engineering disciplines are directly related to specific scientific knowledge: **engineering physics, mathematics and engineering,** and **ceramic, biochemical, materials engineering** are examples.

There is another category of program that can be considered. An engineering degree can be excellent preparation for a career in another discipline, such as law, medicine, or business, to name only a few possibilities. Some engineering graduates change career paths out of circumstance or because their preferences change. Others plan in advance to enter a career other than engineering, but relish the thorough training that an engineering degree provides in analytical problem solving and in personal development. If this is your intention, emphasize breadth of subject matter in your choice of courses, since your breadth of knowledge combined with problem-solving skills will be primary assets in your future.

The above list is not exhaustive, since engineering programs and careers are continually changing and adapting to new knowledge, applications, and societal trends. There is often considerable overlap between disciplines, especially of basic knowledge. In choosing a program, it is always best to investigate the program details in the university calendar, which is usually available via the internet.

4 Becoming a competent engineer

As you enter the first stages of your university engineering program, you may wonder what challenges lie ahead and what you must do to succeed. According to one author [8], the following qualities are needed to be a successful engineer. Your university program is intended to develop these qualities:

- *Mathematical skill and scientific knowledge:* Most of your early courses develop these basic skills, but the real engineering applications of mathematics and science are usually seen later, typically in the later years of your program.

- *Analytical ability:* This is the ability to reduce complex problems into small pieces that can be solved using mathematics and science. You can develop this ability by solving lots of problems—both the hypothetical problems posed in your textbooks, and the real problems that you will face on work-terms or internships.

- *Manufacturing and construction knowledge:* This knowledge is basic but essential, and is obtained through laboratory experiments, projects, work-terms and internships. Sometimes the best lessons occur when an experiment or design does not perform as predicted, and the deviation from theory must be explained. Remember that no project is a failure if something is learned from it.

- *Open-mindedness:* This is the ability to accept unfamiliar ideas from other fields of study. Do not reject unfamiliar ideas before considering them deeply enough to see whether they may help to solve your problem.

- *Decision-making ability:* The ability to make decisions, even in the face of risk or uncertainty, is one of the key abilities of an engineer. Good decisions result when you know the goal that you are trying to achieve and the probabilities of successful outcomes for alternative courses of action. Decision-making ability can be developed with practice.

- *Communication skills:* The ability to express ideas clearly in spoken, written and graphical forms is extremely important, because the engineer typically relies on technologists or technicians to carry out the engineer's ideas. This skill is developed and tested by hundreds of assignments, which you will be asked to create during your university years.

- *Inventiveness:* The ability to think inventively or creatively distinguishes the truly successful engineer from the mundane thinker. This skill is developed only by taking part in projects, contests, writing, design or other activities where creativity is essential.

University courses in engineering attempt to teach you, over a period of years, most of the knowledge and skills listed above. In fact, the last three qualities—decision-making ability, communication skills, and inventiveness—make up about half of this text.

Remember the old saying, that "Employers hire people for two reasons: what they know and who they are." University courses will increase what you know, but only you can determine who you are. Are you a skilful, creative, hard-working problem solver? Can you become a decisive, self-motivated leader and an effective manager? Can you run a meeting, give a speech or negotiate a contract? These qualities and skills partly define who you are, and are as important as what you know.

You can develop and demonstrate "who you are" by your activities outside of the classroom. Are you interested in design contests, student debates, the student newspaper, campus clubs, intramural sports, or volunteering in the local community? Investigate and consider these activities. They will help you to develop management, leadership and communication skills and personal qualities such as strength of character and a positive personality. A word of caution is in order: in budgeting your time, your first priority must be your degree program. However, between academic terms you will have more time available; take advantage of it.

You will make friends during your degree program. You will work hard on challenging problems with them, and some of them will become friends for life. Value them, because they will be part of your community of technical colleagues, who may be as valuable to you in your career as they are before graduation.

5 Challenges for engineering

You will encounter many personal challenges in your education and in your professional life, but you are also a member of a global society, living on a fragile planet. You need not travel far to see examples of inefficient energy consumption, poisoned environments, overtaxed highways, crumbling bridges, or water and air pollution, such as illustrated in Figure 1.3. On a global level, the rapid world population increase, global warming, and crises aggravated by fast communication and transportation of goods, finance and disease, will affect you during your working life. Your goal, as an engineer, must be the efficient and ethical use of resources, as discussed in Chapter 3. Fortunately, societies solve or adapt to these challenges [9], and engineers are inevitably part of the solutions.

The coming challenges will require engineers to participate actively in political and ethical debate. In Canada, engineers are rarely found in the upper levels of government, and are seldom involved in social debate, even when the issues have an engineering content. The influence of engineers on social and political decision-making will increase only if engineers take a greater role in these activities.

Fig. 1.3. Engineering faces many challenges. Billowing smokestacks were once considered signs of prosperity, but now are recognized as evidence of wasteful operations, health risks, and part of the cause of global warming.

More positively, the advent of widespread cheap communication and information processing has ensured that today's engineering students are living in a world of unparalleled opportunity. The information revolution is still in its early stages and it will continue to multiply our strength and intelligence, and help create a world of diversity, freedom and wealth that could not have been imagined by earlier generations. The challenges to the engineer, and the opportunities, have never been greater.

6 Further study

1. Make a list of the "top 10" reasons why you entered engineering.

2. Considering your own personal interests, abilities and potential for development, and referring to the characteristics discussed in Section 4 of this chapter, explain why engineering is your choice as a profession.

3. Should you select a career based on personal interests, aptitudes and previous experiences, or should you select a career based on future job opportunities? (Remember that your previous experience may be very limited and your interests and aptitudes may change as years go by. On the other hand, your knowledge of future job opportunities may be incomplete and job predictions, even when they are accurate, may be affected in future by totally unexpected political, economic or technical developments.) Compare these two methods of career selection. Which is better? Are there other factors that should be considered? How did you make your career choice?

4. If personal characteristics are important in selecting a career or, in particular, a branch of engineering, what are your interests, aptitudes and experiences that led you to select the particular branch of engineering in which you are currently enrolled? How do your characteristics match the characteristics that you think would be necessary in your chosen branch?

5. One of the best sources of information on career selection is an engineer practising in your chosen branch of engineering. If you would like to discuss career choices in detail with a practising engineer, talk to your university professors, your university counselling or placement services, or contact the local chapter of your provincial Association of Professional Engineers.

6. What is the difference between a profession and a job? Your opinions may be influenced by Chapters 2 and 3, which describe the engineering profession.

7. Should the professional person be more concerned about the welfare of the general public than the average person? Should persons in positions of great trust, whose actions could cause great harm to the general public, be required to obey a more strict code of ethics with respect to that trust than the average person? You may find some hints by glancing at Chapters 2 and 3.

8. When you graduate, you will want to know how intense the competition is for career positions. The number of students enrolled in your discipline is a rough indicator of the future supply of job applicants. Do an internet search to find enrollment statistics for engineering disciplines in Canada.

9. The previous question considers the supply of engineers in your year of graduation. The demand in each discipline is harder to predict, but is cyclical in some industries and subject to major trends, such as the computer and communications revolution, the biological gene-modification revolution, and the demographics of birth, death, and immigration. Make a list of at least five possible jobs you might wish to do, and consider how each of them might be affected by current economic trends. Your conclusions cannot be exact, but might be very useful as you seek work-term or internship experience before graduation.

7 References

[1] L. P. Grayson, *The Making of an Engineer*, John Wiley & Sons, Inc., New York, 1993.

[2] C. Lajeunesse, "Engineering," in *The Canadian Encyclopedia*, pp. 703–704. McClelland and Stewart, Toronto, 1988, <http://www.thecanadianencyclopedia.com/> (July 13, 2001).

[3] H. Petroski, *To Engineer is Human: The Role of Failure in Successful Design*, St. Martin's Press, New York, 1985.

[4] S. C. Florman, *The Existential Pleasures of Engineering*, St. Martin's Press, New York, 1976.

[5] Canadian Engineering Accreditation Board, *Accreditation Criteria and Procedures for the Year Ending June 30, 2001*, Canadian Council of Professional Engineers, Ottawa, 2001, <http://www.ccpe.ca/files/report_ceab.pdf> (March 25, 2002).

[6] G. C. Andrews and J. D. Kemper, *Canadian Professional Engineering Practice and Ethics*, Saunders College Canada, Toronto, 2nd edition, 1999.

[7] Industry Canada, *Professional Engineering in Canada: Building a Nation*, Industry Canada, 1998, <http://collections.ic.gc.ca/pec/> (July 12, 2001).

[8] J. R. Dixon, *Design Engineering: Inventiveness, Analysis and Decision-Making*, McGraw-Hill, New York, 1966.

[9] T. Homer-Dixon, *The Ingenuity Gap*, Knopf, New York, 2000.

Chapter

2 The licensed professional engineer

Your university engineering program is not just a set of related courses; it is the entry point into an organized profession with strict admission standards, a code of ethics, and professional regulations. Engineering history in Canada is full of success stories, from the railway across the country to the space program, but it also contains some tragic disasters (see Figure 2.1). This chapter discusses the laws that establish engineering as a profession in Canada and how the provincial Associations admit applicants to the profession. In it, you will learn

- the characteristics of a profession,
- how the engineering profession is regulated by law,
- academic and experience requirements for admission to the engineering profession.

Fig. 2.1. The collapse of the Québec bridge during erection in 1907 [1], and partial collapse in 1916, greatly influenced the laws regulating the engineering profession, which were written in the next decade.

16 Chapter 2 The licensed professional engineer

1 Engineering is a profession

In debating the Professional Engineers Act in the Ontario Legislature [2], engineer H. MacKenzie attributes the following definition of engineering to S. C. Florman, an engineer and prolific writer:

> "A profession is a self-selected, self-disciplined group of individuals who hold themselves out to the public as possessing a special skill derived from training and education and who are prepared to exercise that skill in the interests of others."

Engineering satisfies this definition, as do the older organized professions, such as medicine and law. Engineers possess a high level of skill and knowledge, obtained from lengthy education and experience, and engineering is a creative vocation with a positive purpose.

However, the engineer's working environment differs from that of other professionals. Most engineers are employees of large companies and work in project teams; other professionals are often self-employed and work on a personal basis with clients. The difference is illustrated in the following quote, which emphasizes that the engineering profession is highly regarded, in spite of the employee status of most of its members:

> "The hard fact of the matter is that people need physicians to save their lives, lawyers to save their property and ministers to save their souls. Individuals will probably never have an acute, personal need for an engineer. Thus, engineering as a profession will probably never receive the prestige of its sister professions. Although this may be an unhappy comparison, the engineer should take note that physicians and lawyers both feel that the prestige of their professions has never been lower, and they are mightily concerned; yet [...] engineers are considered to be sober, competent, dedicated, conservative practitioners, without such devastating problems as embezzlement or absconding members and without the constant references to malpractice and incompetence." [3]

2 Regulation of the engineering profession

Since all professions involve skill or knowledge that cannot be evaluated easily by the general public, governments usually impose some form of regulation or licensing on them. The purpose of regulation is to prevent unqualified persons from practising, to set standards of practice that protect the public, and to discipline unscrupulous practitioners.

The United States was the first country to regulate the practice of engineering. The state of Wyoming enacted a law in 1907 as a result of many instances of gross incompetence during a major irrigation project [2]. In the years that followed, all of the United States and all the provinces and territories of Canada enacted licensing laws to regulate the engineering profession and the title of Professional Engineer.

There are three typical methods of government regulation [4]: by direct control through government departments, by establishing independent agencies, and by permitting the professions, themselves, to be self-regulating bodies, which determine standards for admission, professional practice and discipline of their members. In Canada, engineering is self-regulating, as are all of the major professions (medicine, law, accounting, and others). Each provincial and territorial government has passed a law that designates a licensing body, called an Association or Ordre, that regulates the engineering profession. Seven of these Associations also regulate the practice of geoscientists.

In the United States, agencies appointed by the state governments write the regulations and license engineers. In Britain, although chartered institutions promote the practice of engineering and engineering technology for the public benefit, the term *engineer* is not regulated by law. The sign "engineer on duty" is found outside many garages.

2.1 The laws regulating engineering

If your goal, upon graduation, is to practise engineering in Canada, then you must know about the applicable provincial laws. There are 12 Associations (designated with a capital letter, see page 4), that have been delegated the authority, by provincial and territorial law, to regulate the practice of engineering. In 1936, these Associations formed the Canadian Council of Professional Engineers, to set national standards. Each Association, and the Ordre des ingénieurs du Québec, has developed regulations, by-laws and a code of ethics, either of its own or from guidelines published by the CCPE. The provinces have slightly different laws delegating authority to the Associations, and an exhaustive discussion of them cannot be given here, but there are many basic similarities. Membership in an Association entails a licence to practise engineering; Figure 2.2 shows an engineering licence issued in Ontario. The discussion in this chapter emphasizes that province, which has almost half of Canada's engineers.

In Ontario, attempts to regulate engineering began in 1899, but the first legislation for engineering was not passed until 1922. The Professional Engineers Act established a voluntary Association: the Association of Professional Engineers of Ontario (APEO), to regulate engineering. It was called an "open" regulatory Act since it did not require membership in the Association for a person to practise professional engineering; however, in 1937 the legislature amended the Act, and engineering became a "closed" profession. The Act has been amended several times since 1937, most recently in 2001, and as amended [5], can be found in most Ontario public libraries as well as via the internet.

In 1992, the APEO council adopted a revised logo, which appears at the top of the licence in Figure 2.2. The working name was changed to Professional Engineers Ontario, with the acronym PEO. The official legal name is still the

18 Chapter 2 The licensed professional engineer

Fig. 2.2. The engineering profession in Canada is regulated by provincial and territorial law. An engineering licence is shown. (Courtesy of Professional Engineers Ontario)

Association of Professional Engineers of Ontario, but the old acronym (APEO) has been discontinued.

A Regulations document [6] accompanies the Ontario Professional Engineers Act in the official government record. Section 77 of the Regulations contains the Code of Ethics. The Act itself permits PEO to enact by-laws for self-governance. These documents govern several kinds of activities:

- The Regulations govern engineering practice, by providing additional details and interpretation of clauses of the Act. For example, the document defines the academic and experience requirements for admission to PEO, which are stated very generally in the Act.
- The by-laws are rules set up to administer PEO itself. They concern the

meetings of the council, financial statements, committees, and other internal matters.
- The Code of Ethics is a set of rules of personal conduct to guide individual engineers. The Code is discussed in Chapter 3.

Since the regulations and by-laws exist under the authority of an Act of legislature, they have the force of law. Engineers themselves regulate their profession by electing the majority of members to the PEO Council, which also contains members appointed by the Lieutenant Governor-in-Council, and by confirming, by ballot, the by-laws established by the Council. For self-regulation to work effectively, engineers must be willing to participate in Association activities and to serve in the elected positions at chapter, region and council levels.

2.2 The legal definition of engineering

The practice of professional engineering is defined (Professional Engineers Act, Ontario [5]) as:

"...any act of designing, composing, evaluating, advising, reporting, directing, or supervising wherein the safeguarding of life, health, property, or the public welfare is concerned, and that requires the application of engineering principles, but does not include practising as a natural scientist."

In applying this extremely broad definition, the term *engineering principles* is generally interpreted to mean those subjects defined in a university-level engineering curriculum: mathematics, basic science, engineering science, and complementary studies.

The provincial legislative Acts define engineering practice, and place a responsibility on the engineer to ensure that life, health, property, and the public welfare are protected. However, some engineering graduates have jobs that do not appear to carry these responsibilities. Must these graduates be licensed as Professional Engineers? The following advice may be useful.

If the terms of a job truly do not meet the above definition, then it is not really an engineering job, and licensing is not required. However, it is in your interest to be licensed, since the title Professional Engineer and its abbreviation, P.Eng., may not be used by unlicensed graduates, and the lack of a licence may hinder your promotion or prevent future career moves. Moreover, in certain "gray" areas of activity, it is far better to have a licence that is not needed than to break the law by practising illegally.

Beyond the legal requirements for registration and the use of the P.Eng. title on a personal resumé, a licence is a sign of your pride in your profession. Many long-established engineers who are no longer responsible for engineering decisions maintain their membership out of pride in their training, experience, and accomplishments.

3 Admission to the engineering profession

The legislative Acts in the provinces and territories define the admission requirements for the profession. The requirements are similar, but not identical, across the country. To qualify for a licence, an applicant must

- be a citizen of Canada, or have permanent resident status;
- have a minimum age of 18 years;
- satisfy the academic requirements;
- pass the professional practice examination;
- satisfy experience requirements;
- be of good character, as confirmed by referees.

The admission steps for students in accredited university engineering programs are illustrated by Figure 2.3.

Regulations for admission change from time to time. In the recent past, the experience requirements were raised in most provinces from two years to four, with added criteria defining the required quality of the experience. At the same time, an internship process was instituted to help prospective engineers through the process. Consult the information provided by your provincial Association for

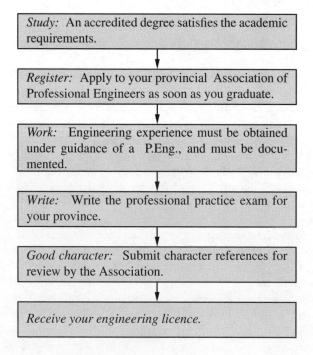

Fig. 2.3. The steps to take to become an engineer via an accredited program.

advice about presenting your personal information, academic qualifications and engineering experience. Most Associations have a student membership designed for easy communication between future engineers and their profession.

3.1 Academic requirements

The provincial Association evaluates each applicant's academic record, and has final authority on all matters related to academic requirements.

The qualifications of each applicant are evaluated according to the requirements given above. The provincial Associations rely on the Canadian Engineering Accreditation Board (CEAB), a standing committee of the CCPE, for advice on engineering programs. This board evaluates engineering degree programs in Canada and the accreditation process in countries where similar systems exist. As a result, graduates of accredited programs in the following countries are treated equally by the respective licensing bodies:

- Canada
- Australia
- Ireland
- New Zealand
- United Kingdom
- United States

Graduates from certain institutions in South Africa, Hong Kong, and France may be granted exemption in Canada from further examinations. For further details about the recognition of programs in these countries, and for a list of programs in other countries that have been listed as "substantially equivalent" to Canadian programs, see [7], which is issued annually. Canadian graduates can expect their academic qualifications to be recognized if they apply for certification to practise as an engineer in the listed countries.

Although persons without a CEAB-accredited engineering degree or equivalent may be admitted to the engineering profession after writing examinations that depend on their background, this is not an easy task, for several reasons:

- There are between 10 and 20 three-hour university-level admission examinations for each branch of engineering. An applicant may be exempt from some subjects, depending on his or her academic record or experience.
- The examination system is not an educational system. Persons applying for the examinations must study on their own and present themselves when they are prepared to write and pass the examinations.
- The provincial Associations do not offer classes, laboratories, or correspondence courses to prepare for the examinations, and there are no tutorial services to review examination results, although some counselling is provided if applicants fail the examinations.

3.2 Experience requirements

To obtain a licence, you must have two to four years of experience, depending on the province. According to the national guidelines developed by the Canadian Engineering Qualifications Board [8], the experience must be at the appropriate level, following the conferring of an engineering degree or the completion of equivalent engineering education. You may register as an Engineer in Training (EIT), Member In Training (MIT), or Junior Engineer (Jr.Eng.), depending on the province. Most Associations have mentorship programs to help you obtain advice from an engineer, who might serve as one of the referees that you will need to vouch for your experience and character. Such programs are especially useful for graduates in work environments where there are few practising engineers.

General requirements: The following is based on the CCPE national guidelines for admission. Work experience acceptable for admission has the following characteristics:

- The experience must be in areas generally satisfying the definition of engineering work.
- The experience is normally in the area of the applicant's degree specialization.
- The experience must be up-to-date.
- The application of theory must be included.
- There should be broad exposure to areas of management and the interaction of engineering with society.
- The experience must demonstrate progression of responsibility.
- Normally, experience is obtained under the guidance, or at least evaluation, of a professional engineer.

In addition to the above criteria, the following factors are considered by PEO; other provinces may differ slightly, and criteria may change:

Canadian experience: At least 12 months of the experience mentioned above must be obtained in a Canadian jurisdiction under the supervision of a person legally authorized to practise professional engineering.

Credit for equivalent experience: A maximum of 12 months of experience, following the conferring of an engineering degree, may be credited for any one of the following:

- experience obtained outside Canada;
- a post-graduate degree or combination of degrees;
- teaching at the level of at least third-year university engineering or the graduating year in a college;
- carefully documented work-term or internship experience obtained after the mid-point of an undergraduate degree;

- military experience.

Documentation: Check the current requirements for the proper method to record the details of your experience. The following data is generally required:

- dates of employment, by month and year;
- names and locations of employers;
- detailed descriptions of the technical responsibilities, identifying the services and products involved; and
- dates of periods of absence from employment.

Experience in a different field: If your experience is not in the field of engineering in which your degree was awarded, the appropriateness of the experience may be evaluated.

3.3 The professional practice examination

Some provincial Associations create their own professional practice examination; others use one that is nationally available. The examination is a three-hour closed-book examination in ethics, professional practice, law, and liability. Every applicant for a licence must successfully complete the examination, regardless of academic qualifications. No exemption is given for similar courses taken in university or elsewhere. Exemption from the examination is permitted only for applicants who are applying to be reinstated after an absence or applicants with at least five years of membership in another Canadian Association.

Time limits: You must submit a licence application before you will be permitted to write the professional practice examination. You may not write the examination until one year after graduation from university, and you must pass the exam not later than two years following the date of submitting the application for membership (or the date of successful completion of all other examination requirements, whichever is later).

Exam schedules: The examinations are generally administered in May, September and December each year. The provincial Association should be contacted for precise locations and dates. These dates and other registration information are published on the web [9].

3.4 Certificates of authorization

In Ontario, the Professional Engineers Act was revised in 1984 to require all engineers or companies providing services to the general public to obtain a Certificate of Authorization from PEO. The key requirements for obtaining a certificate are very simple. Companies must indicate the person who will assume responsibility for supervising engineering work, and must obtain professional liability insurance. If the individual engineer or the company does not have liability insurance, this fact must be disclosed to all clients in writing and clients must acknowledge the disclosure.

4 Provincial engineering organizations

The primary purpose of licensing Associations, Acts, regulations, by-laws, and codes of ethics is to protect the public, not to help engineers. As a result of this public protection, however, confidence in the engineering profession is very high.

In addition to their licensing function, there are fringe benefits from the Associations, such as group insurance, retirement savings plans, employment advice, chapter technical meetings, and social events for members. Occasionally, people who see only these superficial aspects may believe that the licensing bodies are advocacy or service groups for engineers. They are not; they were set up to protect the public by regulating the engineering profession, and the provision of member services is at a low level that does not conflict with the major regulatory purpose.

In most self-regulated professions, there is a two-body organizational structure in which one organization regulates the members of the profession, and an independent organization works on behalf of the members, by setting fees, organizing pension plans, and other activities. A good example is the legal profession; the Law Society in each province regulates members, and the Bar Association in each province works on their behalf. Similarly, in medicine, the College of Physicians and Surgeons regulates medical doctors and the provincial medical association works on behalf of the members. In engineering a similar structure has been proposed. In Ontario, recent changes are resulting in the PEO regulating engineers and the Ontario Society of Professional Engineers working on behalf of employee engineers.

For the most current information on recent developments, academic and experience requirements, and admission procedures, consult the web site of your provincial Association. The addresses of the provincial Associations can be obtained via the CCPE sites [9, 10].

5 Further study

1. Compare the definition of engineer in Chapter 1 with the legal definition of engineering in this chapter. Are the two definitions consistent? Do these definitions agree with your dictionary? If not, what definition must take precedence in your province?

2. What is the goal or purpose of government regulation of engineers? Are the six conditions for admission likely to advance this goal?

3. Find the web site of your provincial Association, and determine the following:

(a) How many years of work experience are required after obtaining an accredited Canadian engineering degree?

(b) Does the regulatory Association indicate how to contact them for internship advice?

(c) Is there a student membership category? If so, how do you become a student member? What are the benefits of student membership?

4. Is there a provincial engineering advocacy association in your province? If so,

(a) Determine and list activities that could, in principle, overlap with the activities of the regulatory Association. Do their terms of reference specify their relationship with the regulatory Association?

(b) Determine the conditions of membership.

(c) Is there a student membership category?

5. Do any provincial Associations have admission requirements additional to those listed in Section 3? If so, what are the requirements, and where are they in force?

6 References

[1] H. Holgate, *Royal Commission: Quebec Bridge Inquiry*, vol. II, Government of Canada, Ottawa, 1908, 7-8 Edward VII, Sessional Paper No. 154.

[2] H. MacKenzie, "Opening address for the debate on the Professional Engineers Act 1968-69, Bill 48," in *Legislature of Ontario Debates, 28th Session, 1969*, vol. 5, p. 4798. Queen's Printer for Ontario, Toronto, 1969.

[3] J. B. Carruthers, "Personal communication to G. C. Andrews," 1978.

[4] H. N. Janisch, "Regulatory process," in *The Canadian Encyclopedia*, p. 1563. Hurtig Publishers Ltd., Edmonton, AB, 1985, <http://www.thecanadianencyclopedia.com/> (April 15, 2002).

[5] Queen's Printer for Ontario, "Professional Engineers Act, R.S.O. c. P-28, as amended," in *Revised Statutes of Ontario*. Publications Ontario, Toronto, 2001.

[6] Queen's Printer for Ontario, "Professional Engineers Act, R.R.O. 1990, Regulation 941, as amended," in *Revised Regulations of Ontario*. Publications Ontario, Toronto, 2000.

26 Chapter 2 The licensed professional engineer

[7] Canadian Engineering Accreditation Board, *Accreditation Criteria and Procedures for the Year Ending June 30, 2001*, Canadian Council of Professional Engineers, Ottawa, 2001, <http://www.ccpe.ca/files/report_ceab.pdf> (March 25, 2002).

[8] Canadian Engineering Qualifications Board, *Guideline on Admission to the Practice of Engineering in Canada*, Canadian Council of Professional Engineers, Ottawa, 2001, <http://www.ccpe.ca/files/guideline_admission_with.pdf> (March 25, 2002).

[9] Canadian Council of Professional Engineers, *P.Eng., The Licence to Engineer*, Canadian Council of Professional Engineers, Ottawa, 2002, <http://www.peng.ca/> (March 25, 2002).

[10] Canadian Council of Professional Engineers, *The Four Steps to Becoming a P.Eng.*, Canadian Council of Professional Engineers, Ottawa, 2002, <http://www.peng.ca/english/students/four.html> (March 25, 2002).

Chapter 3

Professional engineering ethics

The public trusts highly skilled professional people to be accurate, dependable, and ethical. However, opportunities exist in every profession for unscrupulous and self-serving individuals to profit from unprofessional behaviour. News reports of malpractice and misconduct in medicine, law, accounting, and other professions appear occasionally. Such reports are less frequent in engineering, but unprofessional behaviour is equally harmful.

For example, a contractor may attempt to bribe an engineer supervising a construction project into accepting cheap, sub-standard materials or methods. Accepting a bribe is illegal and therefore is clearly unethical, so in this case, the proper course of action is obvious. However, engineers may encounter much more subtle conflicts of interest, where they must choose between two or more difficult courses of action.

In order to guide professional engineers, the provincial regulatory associations publish codes of ethics. These codes vary from province to province, but the basic principles are the same. The scope of this book does not allow discussion of the wording of the codes in all provinces and territories; rather, the PEO Code of Ethics document is analysed as a specific example.

This chapter discusses:

- the PEO Code of Ethics and its legal significance;
- some applications of ethics in the workplace;
- the PEO definition of professional misconduct and the disciplinary powers of the Association;
- the ethical use of engineering computer programs and the engineer's seal; and
- the ethical significance of the iron ring, which is worn proudly by Canadian engineering graduates.

1 Introduction to professional ethics

The Ontario Professional Engineers Act [1] permits PEO to publish a code of ethics [2] to guide professional conduct. The Act further states that one of PEO's additional objects is to establish, maintain and develop standards of professional ethics among its members. To emphasize this fact, applicants for engineering licences are required to write a professional practice examination, in which the Code of Ethics must be applied to hypothetical cases in engineering practice. Some example questions are given at the end of this chapter.

The main purpose of the Code of Ethics is to protect the public from unscrupulous practitioners. However, by enforcing high standards of professional conduct, it raises the public esteem of the profession. The Code of Ethics imposes several duties on the practising engineer, including duties to society in general, to employers, to clients, to colleagues, to the engineering profession, and to oneself.

Occasionally an employee engineer may be faced with the choice between agreeing to an unethical decision or losing his or her job. This situation should not occur, but does from time to time. If the employer is an engineer or a corporation certified by the professional association, then the Code of Ethics applies equally to the employer, and unethical conduct can be reported to the Association. Other situations may require different actions, but an employee engineer should never have to choose between unethical behaviour and a disruption in employment.

The Association staff is available to provide advice and to mediate in cases where an employer is clearly asking an engineer to act unethically. Such cases arise very infrequently, and the first step is to try to solve the problem by discussing it within the company. If that attempt fails and internal solutions are clearly impossible, then a confidential contact may be made with the Association.

1.1 PEO Code of Ethics

Under the Professional Engineers Act, PEO has the responsibility "to establish, maintain and develop standards of professional ethics among its members." Accordingly, PEO prepared and received approval for a code of ethics, recorded as Section 77 in Regulation 941 [2], which accompanies the Act in the official government record. Section 77 is reproduced below, with bold headings inserted for clarity.

General:

1. It is the duty of a practitioner to the public, to the practitioner's employer, to the practitioner's clients, to other members of the practitioner's profession, and to the practitioner to act at all times with,

    i. fairness and loyalty to the practitioner's associates, employers, clients, subordinates and employees,

    ii. fidelity to public needs,

    iii. devotion to high ideals of personal honour and professional integrity,

    iv. knowledge of developments in the area of professional engineering relevant to any services that are undertaken, and

    v. competence in the performance of any professional engineering services that are undertaken.

Duty to society:
2. A practitioner shall,
    i. regard the practitioner's duty to public welfare as paramount,
    ii. endeavour at all times to enhance the public regard for the practitioner's profession by extending the public knowledge thereof and discouraging untrue, unfair or exaggerated statements with respect to professional engineering,
    iii. not express publicly, or while the practitioner is serving as a witness before a court, commission or other tribunal, opinions on professional engineering matters that are not founded on adequate knowledge and honest conviction,
    iv. endeavour to keep the practitioner's licence, temporary licence, limited licence or certificate of authorization, as the case may be, permanently displayed in the practitioner's place of business.

Duty to employers:
3. A practitioner shall act in professional engineering matters for each employer as a faithful agent or trustee and shall regard as confidential information obtained by the practitioner as to the business affairs, technical methods or processes of an employer and avoid or disclose a conflict of interest that might influence the practitioner's actions or judgment.

Duty to clients:
4. A practitioner must disclose immediately to the practitioner's client any interest, direct or indirect, that might be construed as prejudicial in any way to the professional judgment of the practitioner in rendering service to the client.
5. A practitioner who is an employee-engineer and is contracting in the practitioner's own name to perform professional engineering work for other than the practitioner's employer, must provide the practitioner's client with a written statement of the nature of the practitioner's status as an employee and the attendant limitations on the practitioner's services to the client, must satisfy the practitioner that the work will not conflict with the practitioner's duty to the practitioner's employer, and must inform the practitioner's employer of the work.

Duty to colleagues, employees, and subordinates:
6. A practitioner must co-operate in working with other professionals engaged on a project.
7. A practitioner shall,
    i. act towards other practitioners with courtesy and good faith,
    ii. not accept an engagement to review the work of another practitioner for the same employer except with the knowledge of the other practitioner

or except where the connection of the other practitioner with the work has been terminated,
   iii. not maliciously injure the reputation or business of another practitioner,
   iv. not attempt to gain an advantage over other practitioners by paying or accepting a commission in securing professional engineering work, and
   v. give proper credit for engineering work, uphold the principle of adequate compensation for engineering work, provide opportunity for professional development and advancement of the practitioner's associates and subordinates, and extend the effectiveness of the profession through the interchange of engineering information and experience.

Duty to the engineering profession:
8. A practitioner shall maintain the honour and integrity of the practitioner's profession and without fear or favour expose before the proper tribunals unprofessional, dishonest or unethical conduct by any other practitioner.

2 Ethics in the workplace

Everyone wants to work in a fair, creative, productive, and professional environment, and the applicable code of ethics can help to establish such an environment. The PEO Code of Ethics makes several specific statements about the need for fairness, integrity, co-operation, and professional courtesy in paragraphs 1(i), 1(iii), 6, and 7(i). These characteristics are essential in any professional organization, and that is the purpose of such a code: to create a productive and professional workplace.

Although engineering students are not legally bound by the provincial code of ethics, it is a general rule in universities that any behaviour that interferes with the academic activity of others may be classified as an academic offence, which may cause the university's discipline code to be invoked. Academic offences include plagiarism, cheating, threatening behaviour, or other behaviour that has no place in the workplace or in a school for professionals. Therefore, a professional atmosphere is equally important in the engineering student's workplace—the academic environment of classrooms, study rooms, libraries, residences, and the job sites where students work between academic terms. The following examples illustrate the importance of ethical conduct in the student workplace.

Professional behaviour: Engineering students are expected to behave professionally toward employers, colleagues, and subordinates, even though they are not legally governed by a code of ethics. In particular, students on work-terms or internships have an obligation to act professionally on job sites and to keep employers fully informed. The required behaviour includes simple matters such as courtesy, punctuality, and appropriate clothing. For example, hard-hats and steel-toed shoes are essential on a construction site but may be inappropriate in a design office.

As a second example, consider how unprofessional behaviour can undermine the student job-placement process. University placement staff put a great deal of effort into attracting employers to campus. Employers, in turn, invest time and money interviewing students for job openings. If a student accepts a job offer from one employer, whether verbally or in writing, then fails to honour it when a better offer arrives from a second employer, the student would be clearly behaving unprofessionally. If the commitment cannot be honoured because of circumstances beyond the student's control, then there is an obligation to inform the first employer immediately through the placement department and try to minimize the damage. In most cases, an alternative can be negotiated, but only if the student acts promptly and ethically.

Teamwork: As stated earlier, engineering has become a team activity because of the increasing complexity of projects. Engineering students must work effectively with engineers and other professional staff—and the most important goal is the successful completion of the project.

Engineering teams usually include persons with widely different abilities, interests, and education who co-operate by contributing their particular expertise to advance the project. The best engineering team will use each person according to his or her strengths and willingness to contribute to the whole effort. Professionals should not be judged by extraneous factors, such as race, religion, or sex. In addition to being unprofessional and contrary to paragraphs 6 and 7 of the Code of Ethics, such discrimination is contrary to provincial and federal laws on human rights. Within the engineering profession, we must aim for a higher standard of conduct than the minimum set by law. This is the purpose of the Code of Ethics, and is what we would expect of a rational profession like engineering.

Ethical problems: An engineer or an engineering student may occasionally be faced with an ethical dilemma. For example, suppose that you have a colleague or co-worker who has developed a serious personal problem such as drinking, drugs, or mental or emotional unbalance. You are faced with an ethical choice: do you help to conceal the problem and let the individual's health deteriorate, or do you try to get medical or other professional intervention, knowing that this may affect your friend's employment or end your friendship? Both choices are unpalatable.

The solution to an ethical dilemma depends on an evaluation of all factors of the case, using the Code of Ethics as a guide. You might find the design procedure in Chapter 15 of this book to be useful, since it is a problem-solving procedure. As paragraphs 7(i) and 3 of the Code state, an engineer has a duty both to colleagues and to the employer, and this is the root of the dilemma. In view of these conflicting duties, it usually is impossible to give a simple resolution of such a dilemma. The solution requires gathering all of the pertinent information, evaluating the alternative courses of action, and selecting the one that achieves, or comes closest to achieving, the desired goal. When both alternatives are unpleasant, as in the above example, the least undesirable alternative must be chosen.

Resolving disputes: In creative activities such as engineering, differences of opinion are common; in fact, diversity may help to achieve a team's goals. However, disputes must not be allowed to undermine the engineering team, and the best way to settle a dispute is through courteous, direct communication, with the expectation that a full review of the dispute will lead to its solution. In fact, this is the meaning of "good faith" in item 7(i) of the Code of Ethics. This straightforward technique will usually yield a good solution that will be accepted by everyone affected by it.

If you are involved in a dispute, you should never file a formal complaint until all possibilities for personal, informal resolution are exhausted. Sometimes, but very rarely, external agencies or authorities must be contacted to solve ethical problems in the workplace. This action, sometimes called "whistle-blowing," should be the absolute last resort. The professional Association is available for advice in these matters.

3 Professional misconduct and discipline

The duty of the professional Association is to protect the public welfare and to act on complaints from the public. People practising engineering without a licence are prosecuted in the court system. Occasionally it is also necessary to discipline errant engineers. The Association is awarded a wide range of authority to discipline any member who is shown to be incompetent or who is guilty of professional misconduct (see Section 8.(4) of [1]). This authority includes the power to revoke or suspend the licence of a member, as well as the authority, when appropriate, to monitor, inspect or restrict the work done by the engineer.

Although Association staff receive and administer the complaints, a discipline committee, mainly composed of professional engineers, makes the key decisions. Thus, self-regulation of the profession carries through to disciplinary actions as well. In Ontario, the results of disciplinary hearings are published in the PEO *Gazette*, which is circulated to all members as an insert to the magazine Engineering Dimensions.

The most frequently reported complaints concern the Code of Ethics. In the Regulation associated with the PEO Act, breaches of most of the terms of the Code of Ethics are specifically defined to be professional misconduct. In other words, the Code of Ethics says what should be done; the definition of professional misconduct says what must not be done.

Most of the terms of the Code of Ethics are based on common sense and natural justice; therefore, it is not necessary for you to memorize the Code; most engineers find that they follow it intuitively and need never fear charges of professional misconduct. For information, the PEO definition of professional misconduct is reproduced in Section 3.1.

3.1 Definition of professional misconduct

Under the Professional Engineers Act, PEO has the responsibility "to govern its members [...] in order that the public interest may be served and protected." The Act also provides that a member may be guilty of professional misconduct if convicted of a relevant criminal offence, or if found guilty by PEO under Section 72 of Regulation 941, which is reproduced below [3]:

(1) In this section,

"harassment" means engaging in a course of vexatious comment or conduct that is known or ought reasonably to be known as unwelcome and that might reasonably be regarded as interfering in a professional engineering relationship;

"negligence" means an act or an omission in the carrying out of the work of a practitioner that constitutes a failure to maintain the standards that a reasonable and prudent practitioner would maintain in the circumstances.

(2) For the purposes of the Act and this Regulation,

"professional misconduct" means,

(a) negligence,

(b) failure to make reasonable provision for the safeguarding of life, health or property of a person who may be affected by the work for which the practitioner is responsible,

(c) failure to act to correct or report a situation that the practitioner believes may endanger the safety or the welfare of the public,

(d) failure to make responsible provision for complying with applicable statutes, regulations, standards, codes, by-laws and rules in connection with work being undertaken by or under the responsibility of the practitioner,

(e) signing or sealing a final drawing, specification, plan, report or other document not actually prepared or checked by the practitioner,

(f) failure of a practitioner to present clearly to the practitioner's employer the consequences to be expected from a deviation proposed in work, if the professional engineering judgment of the practitioner is overruled by non-technical authority in cases where the practitioner is responsible for the technical adequacy of professional engineering work,

(g) breach of the Act or regulations, other than an action that is solely a breach of the code of ethics,

(h) undertaking work the practitioner is not competent to perform by virtue of the practitioner's training and experience,

(i) failure to make prompt, voluntary and complete disclosure of an interest, direct or indirect, that might in any way be, or be construed as, prejudicial to the professional judgment of the practitioner in rendering service to the public, to an employer or to a client, and in particular, without limiting the generality of the foregoing, carrying out any of the following acts without making such a prior disclosure:

1. Accepting compensation in any form for a particular service from more than one party.
2. Submitting a tender or acting as a contractor in respect of work upon which the practitioner may be performing as a professional engineer.
3. Participating in the supply of material or equipment to be used by the employer or client of the practitioner.
4. Contracting in the practitioner's own right to perform professional engineering services for other than the practitioner's employer.
5. Expressing opinions or making statements concerning matters within the practice of professional engineering of public interest where the opinions or statements are inspired or paid for by other interests,

(j) conduct or an act relevant to the practice of professional engineering that, having regard to all the circumstances, would reasonably be regarded by the engineering profession as disgraceful, dishonourable or unprofessional,

(k) failure by a practitioner to abide by the terms, conditions or limitations of the practitioner's licence, limited licence, temporary licence or certificate,

(l) failure to supply documents or information requested by an investigator acting under section 34 of the Act,

(m) permitting, counselling or assisting a person who is not a practitioner to engage in the practice of professional engineering except as provided for in the Act or the regulations,

(n) harassment.

4 The professional use of computer programs

Computer software is now so essential to all aspects of engineering design, testing, and manufacture that several ethical concerns are created. These include: liability for computer errors, software piracy, plagiarism from the internet, and protection against computer viruses. Each of these topics is discussed briefly.

Liability for computer errors: The engineer is responsible for decisions resulting from his or her use of computer software. In future, this responsibility may be shared, as software development becomes the responsibility of the software engineering profession, but at present, the engineer, not the software developer, is legally responsible for engineering decisions.

Significant engineering programs are never provably defect-free. However, if an engineering failure results from faulty software, legal liability cannot be transferred to the software developer, any more than it can be transferred to an instrument manufacturer because of faulty readings from a voltmeter. To put it even more clearly: *the engineer cannot blame the software* if the engineer makes decisions based on incorrect or misunderstood software output.

In the event of a disaster, the software developer may be liable for the cost of the software, but the engineer will likely be legally liable for the cost of the disaster. Therefore, when software is used to make engineering decisions, the engineer must:

- be competent in the technical area in which the software is being applied,
- know the type of assistance provided by the software,
- know the theory and assumptions upon which it was prepared,
- know the range and limits of its validity, and
- test the software to ensure that it is accurate.

The extent and type of tests will vary, depending on the type of software, but the tests must be thorough. Usually, the engineer must run test examples through the software and compare the output with independent calculations. In some safety-critical applications, computations are performed by two independently written programs and accepted only if the results agree. However, the engineer must never use software blindly without independent validation.

Software piracy: Piracy is the unauthorized copying or use of computer software and is, quite simply, theft. Software piracy occurs more frequently than other types of theft because software can be copied so easily and because the risks of piracy are not very well known.

Software is protected by copyright and trademark laws, as discussed in more detail in Chapter 17. Piracy is an infringement of the developer's rights, and is clearly unethical. Engineers who commit software piracy are subject to civil litigation and, in some cases, criminal charges, as well as professional discipline if the conduct of the engineer is deemed to be "disgraceful, dishonourable or unprofessional" (see Section 72 of PEO Regulation 941).

In addition, it is shortsighted to use unauthorized software in engineering, because an engineer who uses such software runs a much greater risk of legal or disciplinary action if a project runs into problems. In any subsequent legal proceeding, dependence on unauthorized software would be irrefutable evidence of professional misconduct.

Several organizations such as the Canadian Alliance Against Software Theft (CAAST) have been formed to reduce software piracy. CAAST is an alliance of several software publishers, created with the goal of detecting and prosecuting cases of software copyright infringement in Canada. CAAST acts very aggressively through the internet to reduce software piracy by educating the public, by detecting infringers and by enforcing the copyright laws.

The ethical use of software stimulates software development, increases creativity, productivity, and job opportunities. Don't get involved in software piracy!

Plagiarism from the internet: Plagiarism is another ethical problem that has been aggravated by the software explosion. Plagiarism is defined as taking intellectual property, such as words, drawings, photos, artwork, or other creative material that was written or created by others, and passing it off as your own. Plagiarism has become more common in recent years, mainly because of the convenience of cutting and pasting information from the internet. Plagiarism is always unethical and is specifically contrary to several sections in engineering codes of ethics.

Plagiarism could lead to disciplinary action if the actions of the engineer are deemed to be "disgraceful, dishonourable or unprofessional" (see Section 72 of PEO Regulation 941).

Engineering students should be aware that plagiarism can result in severe academic penalties and can delay or abort a promising engineering career. All universities expect students to know what plagiarism means and to avoid it.

When you submit a document with your name on it, such as an engineering report, you take responsibility for everything in the document except for material that is specifically identified as coming from other sources. If material created by others is not clearly and completely cited, then you are liable to be charged with plagiarism.

The format for citing material created by others is explained in Chapter 8 of this book.

Computer viruses: Like other professionals, engineers depend on their computer systems to run effectively and efficiently. The proliferation of computer viruses that damage or overload computer systems is an ever-growing threat. Virus vandalism can cause expensive and dangerous failures. Individuals who participate in creating or disseminating computer viruses are beneath contempt, and every engineer has a duty to expose such conduct.

5 Proper use of the engineer's seal

The provincial Act provides for each professional engineer to have a seal, such as that shown in Figure 3.1, denoting that he or she is licensed. All final drawings, specifications, plans, reports, and other documents involving the practice of professional engineering should bear the signature and seal of the professional engineer who prepared and approved them. This is particularly important for services provided to the general public. The seal has legal significance, since it implies that the documents have been competently prepared. In addition, the seal signifies that a licensed professional engineer has approved the documents for use in construction or manufacture. The seal should not, therefore, be used casually or indiscriminately. In particular, preliminary documents should *not* be sealed. They should be marked "preliminary" or "not for construction."

The seal signifies that the documents have been prepared or approved by the person who sealed them, implying an intimate knowledge and control over the documents or the project to which the documents relate. An engineer who knowingly signs or seals documents that have not been prepared by the engineer or under his or her direct supervision may be guilty of professional misconduct (see item 2(e), page 33). The engineer may also be liable for fraud or negligence if misrepresentation results in damages.

Engineers are sometimes asked to "check" documents, then to sign and seal them. The extent of work needed to check a document properly is not clearly defined, but such a request is usually not ethical. The engineer who prepared the documents or supervised their preparation should seal them. If an engineer did not prepare them, then perhaps the preparer should have been under the supervision of an engineer. The PEO *Gazette*, for example, reports many disciplinary cases involving engineers who improperly checked and sealed documents that later proved to have serious flaws.

Fig. 3.1. A typical professional engineer's seal. On an actual seal, the name of the licensed engineer appears on the crossbar. (Courtesy of Professional Engineers Ontario)

6 The iron ring and the engineering oath

The engineering codes of ethics, such as introduced by the Professional Engineers Act in Ontario about 1948, enjoin each engineer to act in an honest, conscientious manner. However, there is a much older voluntary oath, written by Rudyard Kipling and first used in 1925, called the "Obligation of the Engineer." Those who have taken the oath can usually be identified by the wearing of an iron ring on the small finger of the working hand.

The iron ring is awarded during a solemn ceremony known as the Ritual of the Calling of the Engineer, which is conducted by the Corporation of the Seven Wardens. Although the corporation is not a secret society, it does not seek publicity and is therefore a rather confidential group. It is organized by volunteer engineers and is totally independent of both the provincial professional engineering associations and the universities.

The ceremony is generally conducted during April or May of each year at universities that grant engineering degrees, and it is made known to the graduating students. The Wardens permit non-university engineers to participate, but the ceremony is not usually open to the general public. The iron ring does not indicate that a degree has been awarded, but shows that the wearer has participated in the ceremony and has voluntarily agreed to abide by the oath or "Obligation" of the engineer. The Obligation is brief, but it is a solemn commitment by the graduate to maintain high standards of performance and ethics.

7 Further study

The following questions illustrate ethical dilemmas that are found in engineering practice. Questions 4 to 7 have been reproduced from PEO professional practice examinations. These questions should be answered by citing and explaining the Code of Ethics sections and the Professional Misconduct sections that apply.

1. John Jones is a professional engineer who works in the engineering department for a medium-size Canadian city. He has been assigned to monitor and approve, on behalf of the city, each stage of the construction of a new sewage treatment plant, since he was involved in preparing the specifications for the plant. The contract for construction has been awarded, after a competitive bidding process, to the ACME Construction Company. About 10 days before construction is to begin, he finds a gift-wrapped case of rye whiskey on his doorstep, of approximate value $600. The card attached to the box says, "Looking forward to a good professional relationship," and is signed by the president of ACME Construction. Is it ethical for Jones to accept this gift? If not, what action should he take?

2. Alice Smith is a professional engineer with several years of experience. Ima Turkey, who is a graduate of a school of engineering but has never registered as a professional engineer, approaches her. Turkey offers Smith $1 000 if she will "check" some plans and put her seal and signature on them to make them acceptable to the city official who issues building permits. Is it ethical for Smith to do this?

3. René Brown is a professional engineer who has recently been appointed president of a moderate-size dredging company. Executives of three competing dredging companies approach him. He is asked to cooperate in competitive bidding on dredging contracts advertised by the federal government. If he submits high bids on the next three contracts, then the other companies will submit high bids on the fourth contract and he will be assured of getting it. This proposal sounds good to Brown, since he will be able to plan more effectively, if he is assured of receiving the fourth contract. Is it ethical for Brown to agree to this suggestion? If not, what action should be taken? If Brown agrees to this suggestion, does he run any greater risk than the other executives, assuming that only Brown is a professional engineer?

4. As Chief Engineer of the XYZ Company you interviewed and subsequently hired Mr. A for an engineering position on your staff. During the interview Mr. A spoke of his engineering experience in Québec, where he had worked most recently, and stated that he was "also a member of the Order of Engineers of Quebec." You assumed that he was a licensed Professional Engineer in Ontario. You had business cards printed for his use describing Mr. A as a Professional Engineer, and he accepted and used these cards without comment. Some months later you receive a call from a client, complaining about Mr. A calling himself a Professional Engineer when, in fact, he does not hold an Ontario licence to practise. Upon investigation you found this to be true. You fire Mr. A immediately.

Was your action ethical in this matter? Was A's action ethical or legal? Refer to the Code of Ethics, the Act, and/or the regulation on professional misconduct in your answer.

5. The Professional Engineers Act and the regulations thereunder, which include a code of ethics, govern your conduct as a professional engineer. Discuss the Act and the code as you see them applying to your professional life. Is it necessary to have both an Act and a code? Do they overlap?

6. You are a professional engineer employed by a consulting engineering firm. Your immediate superior is also a professional engineer. You have occasion to check into the details of a recent invoice for work done on a project for which your boss is the project manager, but on which both you and members of your staff have done work.

You are surprised to see how much of your time and particularly the time of one of your senior engineers has been charged to the job. You decide to check further into this by reviewing the pertinent time sheets. The time sheets show that time charged to other work has been deliberately transferred to this job. You try to raise the subject with your boss but are rebuffed. You are quite sure something is wrong but are not sure where to turn. You examine the Code of Ethics for direction. What articles are relevant to this situation? What action must you ethically take?

7. You are an engineer with XYZ Consulting Engineers. You have become aware that your firm sub-contracts nearly all of the work associated with the setup, printing and publishing of reports, including artwork and editing. Your wife has some training along this line and has some free time. You decide to form a company to enter this line of business together with your neighbour and his wife. Your wife will be the president, using her maiden name, and you and your neighbours will be directors.

Since you see opportunities for sub-contract work from your company, you believe that there must be similar opportunities from other consulting firms. You are aware of the existing competition and their rates charged for services and see this as a nice little sideline business. Can you ethically do this and, if so, what steps must you take?

8 References

[1] Queen's Printer for Ontario, "Professional Engineers Act, R.S.O. c. P-28, as amended," in *Revised Statutes of Ontario*. Publications Ontario, Toronto, 2001.

[2] Queen's Printer for Ontario, "Code of Ethics of the Association, Professional Engineers Act, R.R.O. 1990, Regulation 941, Section 77, as amended," in *Revised Regulations of Ontario*. Publications Ontario, Toronto, 2000.

[3] Queen's Printer for Ontario, "Professional Misconduct, Professional Engineers Act, R.R.O. 1990, Regulation 941, Section 72, as amended," in *Revised Regulations of Ontario*. Publications Ontario, Toronto, 2000.

Chapter 4
Engineering societies

Most undergraduates learn about engineering societies through design contests held on campus. For example, the Society of Automotive Engineers (SAE) organizes an annual Formula SAE race (Figure 4.1) for students who design, build, and drive the vehicles. However, the major role of engineering societies in developing and disseminating technical information may go unnoticed by those who are only aware of such contests. This chapter describes:

Fig. 4.1. A Formula SAE race, in which engineering students design, build, and race cars under specified rules. (Photo courtesy of Formula SAE team, University of Waterloo)

- the purpose and history of engineering societies,
- the difference between the roles of engineering societies and the Associations described in previous chapters,
- the importance of engineering societies, and
- some guidance for selecting and joining engineering societies.

1 The purpose of engineering societies

Engineering societies are significantly different from the provincial professional engineering Associations discussed in Chapters 2 and 3. The provincial Associations are created by law and have the legal authority to license engineers and to regulate engineering practice. Membership in (or registration by) the provincial Association is not optional, because a licence is required to practise engineering. However, many voluntary engineering societies have been established to assist engineers. We use the general term *engineering societies* to include organizations that may have widely different names, and that may also be classified as "technical societies," "engineering institutes," or even "learned societies."

Although engineering societies are voluntary, their services are extremely valuable to practising engineers, and usually are too specialized to be provided by the provincial Associations. Engineering societies are usually focussed on a specific engineering discipline, such as computer, electrical, or mechanical, or on a specific industry or specialization, such as nuclear, mining, or manufacturing, or on a specific membership, such as engineering students, engineers in management, or consulting engineers. A list of well-known engineering societies is given later in this chapter.

The purpose of engineering societies may vary, depending on the society, but the main goals usually involve activities of interest to their members, such as:

- publishing technical information,
- developing technical codes and standards,
- encouraging engineering research,
- organizing engineering meetings and conferences,
- organizing engineering design competitions for students,
- organizing short courses on specialized topics, and
- advocating on behalf of their members.

These activities have an immense impact on the engineering profession. When engineering societies encourage research and publish the results, they stimulate the creation of knowledge and innovative new products. Engineering societies are also a major developer of codes and standards that guide engineering design to assure safety and product quality. The meetings, conferences and short courses organized by engineering societies are extremely useful for the rapid dissemination of new ideas. Everyone benefits from this free exchange of valuable information.

In addition, all engineering societies advocate on behalf of their members, and this is the most distinctive difference from the provincial professional engineering Associations. The primary mandate of the provincial Associations is to protect the public, so the Associations would have a conflict of interest if they took a leading role in advocating for engineers.

The role of engineering societies varies among countries. In Britain, for example, the long-established technical institutions assess education and training programs, as well as the qualifications of individuals for registration as "Chartered Engineers" by the Engineering Council, a co-ordinating and regulatory body. The technical institutions are also represented in the Engineering and Technology Board, an advocacy body. In the United States, many technical societies publish codes of ethics, whereas in Canada, codes of ethics are enforced by the provincial Associations.

2 The history of engineering societies

Britain led the way in establishing engineering societies, with the Institute of Civil Engineers in 1818, followed by the Institution of Mechanical Engineers in 1848.

In the decade following 1848, many additional societies were established [1].

In the United States, the first engineering society was the American Society of Civil Engineers, founded in 1852. Several others were established thereafter, for example, the American Society of Mechanical Engineers in 1880 and the American Society of Heating and Ventilating Engineers in 1894.

The American Institute of Electrical Engineers was established in 1884 as a national society, primarily for electric power engineers, but later electronics and communications engineers were included. In 1912, the Institute of Radio Engineers was formed as an international body for both professionals and non-professionals in the new field of radio communications. In 1963, these two organizations merged to create the Institute of Electrical and Electronic Engineers (IEEE), which is now the world's largest engineering society. The IEEE has members in all parts of the world. Within the IEEE, there are over 40 professional-interest societies and councils, from Aerospace and Electronic Systems to Vehicular Technology.

Several engineering societies were formed in Canada shortly after Confederation, between 1867 and 1900. Examples are the Canadian Institute of Surveying in 1882 and the Engineering Institute of Canada in 1887, originally the Canadian Society of Civil Engineers. The Canadian Institute of Mining and Metallurgy was formed in 1898 [2]. The formation of student societies began in 1885 with the Engineering Society of the University of Toronto; surprisingly, the "Society was, indeed, a 'learned society,' and published and disseminated technical information [...] in addition to looking after the University undergraduates in engineering." [1]

In keeping with trends toward greater specialization, several Canadian engineering societies have been recently re-organized and new societies created. The Engineering Institute of Canada (EIC), which had served all disciplines for many years, recognized that it could not serve the many diverse specialties of engineering within a single organization; the EIC is now a federation of member societies, as listed in Section 4.

3 The importance of engineering societies

When you leave the university and move into engineering practice, you will want to undertake the sometimes difficult task of keeping your technical competence up-to-date. Help is available from the engineering societies, which are, in reality, "learned" societies, with a vast storehouse of useful knowledge.

Figure 4.2 shows the relationships between the practising engineer and the agencies that regulate and assist engineers. In each province, the government has passed a law to regulate the engineering profession, and under the law, an Association has been created. For example, in Ontario, the Ontario government passed the Professional Engineers Act, which created Professional Engineers Ontario (PEO) to license engineers and regulate the engineering profession, as discussed

44 Chapter 4 Engineering societies

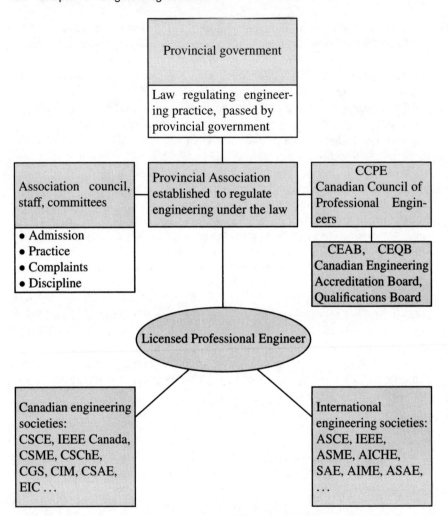

Fig. 4.2. The relationship of the licensed engineer to the provincial engineering Association, CCPE, and Canadian and international engineering societies.

in Chapter 2. The council, staff and committees of each Association enforce the relevant Act, and admit, regulate and discipline members of the profession.

The provincial Associations act jointly through the Canadian Council of Professional Engineers for matters of national and international scope. Two standing committees of the CCPE are the Canadian Engineering Accreditation Board (CEAB), which advises the Associations on the quality of engineering degree programs, and the Canadian Engineering Qualifications Board (CEQB), which advises the Associations on standards for practice, admission to the profession, and ethical conduct.

The CEAB evaluates new engineering programs during the year prior to the graduation of the first graduates, and regularly revisits established programs to verify that minimum standards are met. Total denial of accreditation occurs rarely, but a significant minority of programs are accredited for less than the full term of six years, with renewal requiring action by the university. The list of accredited programs, published annually, does not distinguish between limited and full-term accreditation.

Although the provincial Associations regulate the engineering profession, they cannot provide the lifelong learning that all engineers require to keep their knowledge up-to-date. The simplest way for you to keep up with the rapid changes in your profession is to join and participate in the activities of an engineering society specializing in your discipline. Most practising engineers are members of both a Canadian engineering society and an international engineering society.

Several Canadian societies exist to advocate for engineers in a more personal way. Although the members of these societies are licensed engineers or engineering students, the societies are not, strictly speaking, "learned" societies. These advocacy societies focus on improving the welfare of the individual, rather than the advancement of a specific discipline. Examples of such societies are the Canadian Federation of Engineering Students and the Canadian Society of Professional Engineers (CSPE). The CSPE is an umbrella body for provincial societies that promote the welfare of licensed engineers. The first provincial advocacy society to be formed is the Ontario Society of Professional Engineers, which describes itself as "a member-interest society for Ontario's professional engineers." The National Society for Professional Engineers (NSPE) is an advocacy group for U.S. engineers.

Finally, honorary engineering societies exist, such as the Canadian Academy of Engineering, which was established in 1987. The Academy exists to promote engineering, to recognize important service, and to speak as an independent voice for engineering. It cannot be joined; the Fellows of the Academy elect new members, from all disciplines, based on their record of distinguished service and contribution to society, to Canada, and to the engineering profession. The Academy is distinguished and exclusive; membership is limited to a maximum of 250 Fellows.

4 Choosing your engineering society

As a professional engineer in a rapidly changing world, you have an obligation to maintain your competence. Engineering societies are one of the best sources of up-to-date technical information, and the societies are as important today as they were during the Industrial Revolution. Each professional engineer should be a member of at least one society. For tax purposes, engineering society dues are deductible from personal income, for practising engineers, under Canadian income tax laws.

Chapter 4 Engineering societies

Your choice of society depends mainly on your engineering discipline. Most major societies have student chapters, and you should, if possible, participate in their activities to learn about the society. However, if student chapters do not exist in your area, then you can learn about engineering societies easily through the internet or by speaking with a practising engineer. Internet addresses are listed below for the larger societies, and a good publication that summarizes the activities of over 450 societies throughout the world is the *Directory of Engineering Societies and Related Organizations*, published annually by the U.S. Engineers Joint Council [3]. You should find it in your university library. This directory lists the purpose, membership, address, dues and many other statistics for each society.

4.1 Canadian engineering societies

Engineering Institute of Canada (EIC): As mentioned previously, the long-established EIC is now a federation of engineering member societies that co-operate to advance their common interests, such as maintaining engineering competence through continuing education courses, recognizing engineers by special awards, preserving Canadian engineering history, and creating opportunities for Canadian engineers to participate in disaster relief activities. The EIC lists the following member societies:

- Canadian Geotechnical Society (CGS),
- Canadian Medical and Biological Engineering Society,
- Canadian Society for Chemical Engineering (CSChE),
- Canadian Society for Civil Engineering (CSCE),
- Canadian Society for Mechanical Engineering (CSME),
- Canadian Nuclear Society,
- Canadian Society for Engineering Management,
- IEEE Canada: the Institute of Electrical and Electronics Engineers, Inc.,
- Canadian Maritime Section of the Marine Technology Society.

All of the member societies have web sites that can be reached by web search, from the EIC web site <http://www.eic-ici.ca/> or from the CCPE web site <http://www.ccpe.ca/>. These sites and those mentioned below were current at the time of publication of this book.

Canadian engineering societies that serve more specialized interests include the following:

- Canadian Institute of Mining and Metallurgy (CIM) <http://www.cim.org/>,
- Canadian Society of Agricultural Engineering (CSAE) <http://www.csae-scgr.ca/>.

Section 4 Choosing your engineering society   47

Canadian advocacy societies:
- Canadian Federation of Engineering Students (CFES) <http://www.cfes.ca/>,
- Canadian Society of Professional Engineers (CSPE) <http://www.cspe.ca/>,
- Ontario Society of Professional Engineers <http://www.ospe.on.ca/>,
- Association of Consulting Engineers of Canada (ACEC) <http://www.acec.ca/>.

Canadian honorary society: The Canadian Academy of Engineering <http://www.acad-eng-gen.ca/>.

4.2 American and international engineering societies

In 1904, five U.S. societies co-operated to create the United Engineering Trustees, Inc. The five founding societies have undergone changes over the years, and in 1998 the United Engineering Trustees, Inc. was reorganized as the United Engineering Foundation, Inc. The founding societies, or their successors, have expanded their roles to include engineers everywhere and are listed below:

- American Society of Civil Engineers (ASCE),
- American Society of Mechanical Engineers (ASME),
- The Institute of Electrical and Electronic Engineers (IEEE),
- American Institute of Chemical Engineers (AIChE),
- American Institute of Mining, Metallurgical and Petroleum Engineers (AIME), a corporation owned by its four member societies:
  - Society for Mining, Metallurgy, and Exploration (SME),
  - The Minerals, Metals & Materials Society (TMS),
  - The Iron and Steel Society (ISS), and
  - The Society of Petroleum Engineers (SPE).

The U.S. societies listed above have web sites that may be reached directly by web search from the United Engineering Foundation, <http://www.uefoundation.org/>, or from the AIME web site, <http://www.aimeny.org/>.

Hundreds of U.S. and other engineering societies exist, for the purpose of developing and disseminating information about specific interests. These societies are too numerous to list here, but can be found by web search. Some examples are:

- Society of Automotive Engineers (SAE) <http://www.sae.org/>,
- American Gear Manufacturers Association (AGMA) <http://www.agma.org/>,
- American Society for Engineering Education (ASEE) <http://www.asee.org/>.

A principal American advocacy association is the National Society for Professional Engineers (NSPE), <http://www.nspe.org/>.

5 Further study

1. One of the main roles of engineering societies is the dissemination of information, usually through periodicals. Visit your engineering library reading room and make a list of the top 10 technical periodicals or technical magazines of greatest interest to you, including transactions, journals and popular magazines. Are any of them published by engineering societies? If so, find and read the web site for the engineering society that publishes the periodical. If not, explain why your engineering library does not include technical periodicals from an engineering society of interest to you.

2. Consulting engineers have an advocacy group, the Association of Consulting Engineers of Canada. What is the role of ACEC and how does it differ from CSPE and Ontario Society of Professional Engineers?

3. Using the internet, find at least five student design competitions presently underway or held in the last five years, such as the Canadian Engineering Competition sponsored by CFES or the Formula SAE race sponsored by SAE.

4. Using the internet, identify the web sites of at least 10 engineering societies that might be relevant to your interests but are not listed in this chapter. *Hint:* search by discipline, activity, manufacturing area, technical activity, or other specialized characteristic.

5. Using the internet, determine in more detail the purpose (mission), activities, criteria for admission, and financial support of the Canadian Academy of Engineering. The web address is listed above.

6 References

[1] L. C. Sentance, "History and development of technical and professional societies," *Engineering Digest*, vol. 18, no. 7, pp. 73–74, 1972.

[2] R. B. Land, "Associations," in *The Canadian Encyclopedia*, pp. 109–110. Hurtig Publishers Ltd., Edmonton, AB, 1985, <http://www.thecanadianencyclopedia.com/> (April 15, 2002).

[3] Engineers Joint Council, *Directory of Engineering Societies and Related Organizations*, Engineers Joint Council, New York, 2001.

Chapter 5

Advice on studying and exams

You have spent much time, money and effort to enter an engineering program in an unfamiliar environment. This chapter will help you to succeed or excel academically. In fact, for best results, you should "read it before you need it." You will learn:

- how much time you should expect to spend studying,
- how to work effectively,
- how to prepare for examinations, and
- what to do if things should go wrong.

If you follow this advice (see Figure 5.1), you will almost certainly succeed and still have enough free time to enjoy the diversity of university life. However, only the main strategies for university success are covered here; for more written help, consult references such as [1] and [2] in a library. For more personal advice, speak to someone; professors and senior students are usually happy to pass on their experience, and the staff of your university counselling service provides personal and confidential help.

Fig. 5.1. Resist the urge to sit in the rear rows of large university lecture halls. Sit in the front rows and participate.

1 The good and bad news about university studies

The good news is that you have survived a very competitive engineering admission process, so you almost certainly have the academic ability to graduate. The bad news is that your high-school study skills probably will not be good enough for university. Professors report that in some programs, the average grade for a typical first-year engineering student is significantly below the student's high-school average. The explanation for this phenomenon is that university programs typically admit only the top fraction of high-school graduates, so almost everyone in your university class was a top student in high school. If you want to cope with the higher university standards, you must study effectively.

Moreover, engineering courses are difficult. This should not be surprising; courses are difficult in all professional schools, including law, medicine, optometry, accounting, and others. Professional schools cannot risk losing their accreditation because of low standards. However, there is no required quota of student failures, as some students may believe, and the greatest problem for students is not high standards; it is coping with new-found freedoms.

2 How much study time is required?

An informal survey of university students showed that, in addition to lectures and laboratories, students typically spent 28 to 32 hours per week completing assignments and studying. That is 5 hours per day for a six-day week! You might need a little more or a little less study time, depending on your capabilities. In any case, organize your timetable so that, after time is allotted for lectures, clubs, sports and entertainment, about 30 hours per week is free for assigned work and studying.

At a university, free time tends to get lost! Universities generally treat students as responsible adults. Some students may be tempted simply to relax and to neglect their studies, with tragic consequences. There are no watchful parents or high-school teachers to worry about, and university professors do not check attendance or remind students to submit assignments. However, students who lose sight of academic priorities get a rude awakening when they learn that universities promote students only on the basis of demonstrated performance, not on future potential. Don't be tempted to skip classes, ignore assignments or procrastinate.

3 Preparing for the start of lectures

A few simple tasks can get your term off to a good start:

- *Prepare a timetable*—As mentioned above, schedule your lectures, clubs, sports and entertainment, but leave about 30 hours per week free for reading and studying.

- *Buy the specified course textbooks*—Many university bookstores encourage textbook orders by internet or telephone, so that bookstore lineups may be avoided. Get the texts as soon as possible; sometimes texts sell out and must be re-stocked. Used books might be available, but make sure that you obtain the correct edition. Do not sell your textbooks; a book that you have read thoroughly is even more valuable when you refer to it later.

- *Skim through the texts before lectures start*—As soon as you get your texts, skim through them. If you read even a little bit about a subject before a lecture, you will be astounded how simple and logical it becomes. If you cannot do this for every subject, choose only the most difficult courses. You will be able to ask good questions, clear up doubts, and avoid panic at the end of the term.

- *Get a detailed outline for each course*—Each of your professors will probably distribute a detailed course outline in the first lecture or make it available by web browser. If no outline is provided, speak to the professor; a clear understanding of the course content, marking scheme, and required assignments is essential at the beginning of the term.

4 A checklist of good study skills

The checklist below lists 10 basic points; they are common sense, but are worth repeating. The checklist is arranged in order of importance, with the most important rules first. The first 6 points are absolutely essential—if you are not following all of them, you will likely have academic problems:

1. *Attend lectures regularly, and pay attention*—This is an extremely important rule; a few skipped lectures may seem innocuous, but in a fast-moving course, they can begin a vicious cycle of losing interest and skipping more lectures, which ends when the entire class is dropped.

2. *Submit all assignments*—Even if assignments are not counted as part of the term grade, the final exam will undoubtedly include topics from the assignments. It is usually advisable to work on assignments in tutorial sessions where you have the help of a teaching assistant.

3. *Get a regular place to study*—If your residence is not suitable for studying, use the study tables in the university library or unused classrooms.

4. *Don't overload yourself with part-time employment*—Outside work is a common cause of poor performance. You may have had a part-time job in high school, but time is much more limited in university. Speak to a university student-loan counsellor if you have financial problems.

5. *Don't procrastinate until a deadline arrives*—Major assignments can turn into disasters if left to the last minute. Even an hour spent early in the project, organizing and scheduling how and when you will complete the assignment, will pay off dramatically as the deadline gets closer.

6. *Don't schedule every weekend for trips, parties or travel home*—Some weekend work is essential, whether it is studying, catching up, reviewing, writing reports, or preparing for examinations.

7. *Don't stay up all night*—Late nights, whether for work or play, disrupt your life and can lead to serious health problems. To study effectively, you need a minimum of 7 to 8 hours sleep each night, or at least 50 hours per week.

8. *Concentrate on your work*—When you study, do it! Work will be done more quickly if you concentrate on it. Don't study in bed or try to do two things at once, such as watching TV with your books open; you're just wasting time.

9. *Read ahead*—A brief preview of your texts every week gives you an overview of each subject, and greatly aids the learning process. If you cannot do this for every course, just preview your most difficult courses. If a recommended course text is not very readable, ask your professor, librarian or bookstore to suggest an alternative text, or seek information on the internet.
10. *Reward yourself when you study*—When you are studying, take a 5- to 10-minute break every hour. Stretch, walk, exercise or reward yourself with coffee or a snack. If you have a deadline for a major project, promise to reward yourself with an entertaining night off when the deadline is met.

5 Developing a note-taking strategy

Note-taking during lectures is a personal matter and most students have their own preferences. However, the two extremes are usually both bad: writing down nothing and writing every word. Before the term starts, you should decide what your note-taking strategy will be for each course. Typical note-taking strategies are as follows:

- *Minimal*—Your class notes are just headings intended to remind you what topics were covered. You plan to study the course using the textbook or other course aids. This strategy may be suitable in a course where printed course notes are available. However, if you follow this strategy, write at least one page for every lecture, to record information, such as deadlines, that may not be in the printed notes. Even if a lecture is cancelled, it is reassuring to have a sheet of paper to remind you of this fact.
- *Moderate*—Your class notes are more thorough, including headings, derivations, and quotations that will be reviewed and compared with the textbook. This strategy is usually suitable for most courses, provided that a textbook has been specified and the professor is following it. For this strategy, the general rules are:
  - Make notes for every lecture;
  - Include the course number and date on each page of notes;
  - Make your notes brief but complete, with appropriate headings, textbook page references, the topics discussed, derivations that are difficult and points that are controversial.
- *Maximal*—Your class notes are extremely thorough. Perhaps you add to them by discussion with other students. This strategy is usually necessary when your notes will be your main source of study because no textbook or other course aids are available.

Regardless of your note-taking strategy, you should organize your notes in a file or binder, and review them, along with the textbook or other course aids, to ensure that they agree. If you have questions, raise them at the next lecture or tutorial session. Reviewing your lecture notes for a few minutes before the next lecture will greatly improve your understanding of the material.

6 Hints for assigned work

Most assigned problems are intended to apply the principles introduced in lectures, and you should submit them regularly, on time. However, you must avoid the extremes of spending too little or too much time on them. An engineer must strive for an optimum return for the time invested. Occasionally, enthusiastic students spend an excessive amount of time on interesting projects. Remember the "law of diminishing returns": as the time spent on a project increases, the incremental benefit decreases. This saying applies to what you will learn as well as to the grade you will earn.

Many students collaborate when working on assignments, and this joint work is usually beneficial. However, there is a fine line between collaboration and copying. When you submit an assignment with your name on it, it is assumed that you have prepared all of the material in the assignment, except where you indicate that material has been taken from other sources and the sources are cited. To put this point in clearer terms: an exchange of ideas by talking to one another is collaboration; an exchange of written materials is copying.

Another word for copying is *plagiarism*, which is defined as taking any intellectual property, such as words, drawings, photos, or artwork, that was written or created by others and presenting it as your own. If you submit an assignment with your name on the front, and include material inside that has been taken from others but is not cited, then you have committed plagiarism, and you may be subjected to severe disciplinary action, including dismissal from the university. Plagiarism is discussed at several places in this book; in particular, the proper method of using material provided by others and citing the source is described in Chapter 8.

7 Preparing for examinations

Few people like exams, but they are essential in the university. Exams were originally devised, centuries ago, to prevent favouritism, and they are used today for the same purpose. Exams ensure that students are promoted on knowledge and ability, and not on apple-polishing, bribery, or luck. There are no limits on the number of students who can pass. Exams, therefore, are merely an impartial metre-stick, applied to see that everyone measures up. Try to view them in a positive way, as an achievable challenge.

Examinations are also a learning experience; in fact, the effort put into summarizing and organizing the course material in preparation for an exam is usually very efficient learning.

As soon as the exam schedule is known, usually a few weeks before the end of the term, you should begin your exam routine, as described in the following paragraphs.

Make up a personal schedule. Include the scheduled exam dates, the remaining lectures, and any important social events, interviews, or other obligations. Identify the uncommitted days remaining until exams begin. This is the time that you can control.

Plan time to review each subject before the exams begin. If your assignments interfere with this study schedule, talk to your professors; maybe an assignment can be delayed or shortened.

Collect your notes and organize them. At this point you will appreciate having a course outline and dated notes for each lecture. Old exams are also very useful references; obtain them and list the topics covered by the old exams. Where are these topics in your course outline?

Review each course systematically. Ideally, you should review a course in three stages: first, review the outline and purpose of the course; second, reacquaint yourself with the main topics; and, third, review derivations, assignments and problems from previous exams. Prepare brief review notes as you go, containing definitions, summaries and lists of equations. These notes will be very valuable for a final review, the day before the exam.

Avoid the use of mood-altering drugs of any kind. Even coffee and alcohol, in excess, can cause trouble. Collaborating with fellow students is useful, but avoid "all-nighters."

Save the afternoon before the exam for a final review. You should have started studying early enough that you can spend these last few hours re-reading your notes, trying old exam problems and reviewing key points. Then relax, and make sure you get a good sleep.

8 Writing examinations

Even if you are well prepared, it is normal to feel slightly tense before an exam. Don't let it bother you; everyone else feels the same, even if you can't see it. These suggestions may help you:

- Take a brisk walk before the exam. The mild exercise helps to combat anxiety and clear your mind.
- Arrive a little early; make sure you have writing instruments, you visit the toilet, and are prepared.
- *Read the exam paper!* It is amazing how many students waste time giving excellent answers to questions that were not asked.
- Always solve the easiest question first. This builds confidence and clears your mind for more difficult questions.

- If you are faced with a really tough question, read it thoroughly and go on to the next question. Your mind will work on it subconsciously and when you come back to it, you may have the answer.
- Write clearly and solve problems in a logical order. This shows a methodical approach to problem-solving, and almost always will get you a higher grade.
- If you are short of time, write down how you would solve the problem if you had more time. This explanation is better than nothing. Remember that the exam is a communication between you and your professor, so you may include any comments, references or explanations that you would make orally.

9 When things go wrong

Everyone has bad luck, occasionally: sickness, a car accident, family problems, legal problems, social or other problems. Every university has a procedure for helping students with serious personal problems at exam time. However, the student must take the initiative! If you have a serious problem that clearly interferes with your ability to write an examination, then tell someone! The appropriate person may be your medical doctor, counsellor, professor, department chair or the exam-room supervisor (proctor). However, the earlier you speak out, the easier it is to remedy the problem. For example, don't wait until you receive your exam results to say that you were seriously ill.

After the examination, if you feel that there is an error in your grade, it can be reviewed. This causes some inconvenience, and should not be done casually. A formal letter to the university Registrar explaining your reasoning (and there must be a reason) will set the process in motion. Refer to your university calendar for more information about the examination review process.

Finally, if you find that the subjects you are studying are not really the topics that you expected when you enrolled in engineering, you may have enrolled in the wrong program. Your professors and faculty counsellor may be able to help you define your career objectives more clearly. This textbook, particularly Chapters 1, 2, 3, and 4, is intended to help you clarify your career choice. However, if you still have doubts, get more advice and guidance. Every university has a counselling service available to assist students, such as shown in reference [3]. If you are not sure how to contact your university counselling service, try searching your university web site.

10 Further study

1. Using your computer, prepare a calendar for the end of your current academic term, including an end-of-term exam schedule, as suggested in Section 7 of this chapter. Insert dates of known university or class events. Compare and exchange your template with other class members to ensure that all essential events are included. When you receive your final exam schedule, insert the exam dates as suggested in Section 7, and schedule your exam review.

2. Using the internet, search for entries including the word *plagiarism*. Search also for entries under the heading "Academic Offence." What is your university's policy on plagiarism? Does your university classify plagiarism as an academic offence?

3. Using the internet, search your university web site for entries including the phrase "study skills." If you cannot find this entry on your university web site, try reference [3]. Compare the study skills on the web site with the suggestions in this chapter. Do the suggestions that you found on the internet substantially agree or disagree with this chapter?

11 References

[1] D. H. O'Day, *How to Succeed at University*, Canadian Scholars' Press, Toronto, 1990.

[2] M. N. Browne and S. M. Keeley, *Striving for Excellence in College: Tips for Active Learning*, Prentice Hall, Upper Saddle River, NJ, 2001.

[3] Counselling Services, *Study Skills Package*, University of Waterloo, Waterloo, ON, 2002, <http://www.adm.uwaterloo.ca/infocs/> (April 20, 2002).

Part II
Engineering communications

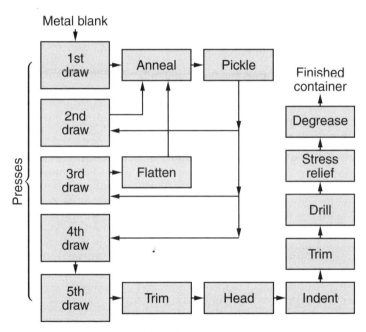

Fig. II.1. Flowcharts are used in all engineering disciplines to communicate the essentials of a sequence of steps or events. The example above shows the steps in the manufacture of seamless metal containers by a process called deep-draw forming.

Engineers are creative people! However, creative ideas must be conveyed clearly to others. Engineers need effective communication skills in order to inform and persuade people in the course of a project and to record the progress and results. In addition to normal business documents, engineers must produce clear and correct engineering reports, drawings, diagrams, graphs, letters, memoranda, and electronic communications.

Engineers must also interact with the business world, and the ability to write and speak effectively in business transactions is important. On rare occasions, you may have to defend your actions and conclusions to business associates or even under the scrutiny of a lawyer in court. Most engineers seek to avoid legal confrontations, but some engineers who are effective communicators regularly act as expert witnesses.

Effective communication skills equip you for a successful career, and the time you spend mastering these skills will be rewarded many times over in future years. Part II explains the basics of effective written communication, in the following four chapters:

Chapter 6—Technical documents: What types of documents do engineers use? This chapter describes 10 basic documents that are typically found in an engineering project or are common to engineering and business, and explains briefly how to prepare them.

Chapter 7—Technical writing basics: Are you familiar with the basic rules of English grammar? This chapter gives you hints on writing style and punctuation and discusses some errors to avoid. The chapter concludes with hints for the use of the Greek alphabet in technical writing and a grammar self-test so you can test your ability to write English.

Chapter 8—Formal technical reports: This chapter concentrates on the technical report, which is one of the most common engineering documents, and one of the most important. Technical reports may discuss a wide range of topics, and are normally organized according to the basic rules explained in this chapter. Knowing how to begin is also important in report writing, and this chapter explains six simple steps in writing an engineering report. The chapter concludes with a report checklist to help you review your report before submitting it.

Chapter 9—Report graphics: The term *graphics* includes all of the diagrams, charts, graphs, sketches, artwork, and even engineering calculations. Graphics are especially important in engineering work, where you frequently must describe trends, patterns, geometrical concepts or complex shapes. This chapter describes basic principles of creating good graphics and the rules for including them in documents such as technical reports.

Chapter 6

Technical documents

Technical writing is distinguished by an emphasis on objectivity, clarity, and accuracy. Graphics are included as necessary to explain and clarify. The information must be presented logically and objectively, and all conclusions must be supported by the document contents or by reference to other documents. Considerable skill and creativity may be required to present complex material clearly and concisely, but artistic licence is not allowed when presenting conclusions.

Technical writing is an art, and requires talent and intelligence, but it is also a craft, the tools (see Figure 6.1) and strategies of which can be learned. In this chapter, you will learn about the kinds of documents you may have to write as an engineer and acceptable formats for them.

There are many reference books containing advice for writing technical documents; a selection is given by references [1–9] at the end of this chapter.

Fig. 6.1. The computer has done to the typewriter what the automobile did to the horse and buggy. In the recent past, an engineer would instruct a typist equipped with a modern version of the antique shown, but now engineers are expected to participate in computerized publishing. Higher-quality documents result, but expectations of quality are correspondingly higher.

A technical document, whether produced at work or for academic credit, is intellectual property and subject to copyright, as discussed in Chapter 17.

1 Types of technical documents

The range of technical documents produced by engineers is as large as the set of organizations and situations in which they find themselves. Nevertheless, the most common document types are easy to list. Most technical documents are modified forms of the documents described here.

Follow the general guidelines in the following sections, unless your organization has specific style and format rules for technical documents.

1.1 Letters

Letters are used to communicate with people outside the company or organization of the sender. The distinguishing characteristic of a technical letter is not found in its format, but in the requirements for clarity and objectivity.

A typical letter, shown in Figure 6.2, has eight basic parts:

1. company letterhead including the sender's return address,
2. date that the letter was signed,
3. inside address of the recipient,
4. subject line,
5. salutation,
6. body, organized as required by the content, but beginning with an introduction that presents the purpose of the letter and desired outcome if any, and ending with a closing statement of follow-up actions desired or to be taken,
7. complimentary closing,
8. signature of sender with printed name and title.

When attachments are included with the letter, a note is made at the bottom of the page, as shown in the example.

Format the date as shown in the figure, unless you are required to do otherwise. Never use the format 09/12/2002, which is ambiguous, and typically means 12 September 2002 in the United States, but 9 December 2002 in Europe.

1.2 Memos

The word *memo* is a short form of *memorandum,* and literally means "something to be remembered." However, normally the word simply means a written communication between people within a company or organization. Memos are used for all informal and formal written communication, except that letters may be preferred for very formal internal communication such as the establishment of a contractual obligation.

The memo has seven parts, as shown in Figure 6.3:

1. heading: the title Memorandum or Memo,
2. Date line,
3. To line: name of recipient,
4. From line: name of sender,
5. Subject line: subject of memo,
6. body,

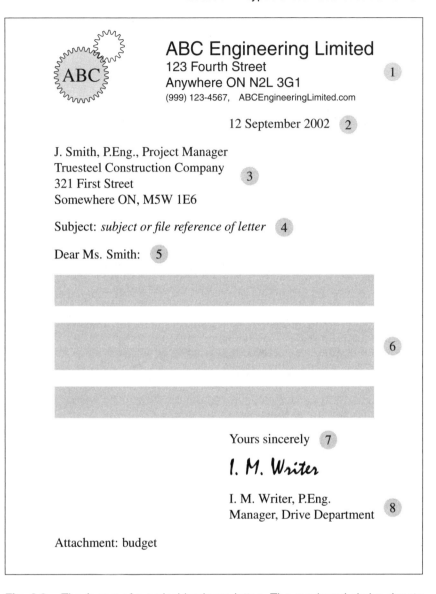

Fig. 6.2. The format of a typical business letter. The numbered circles denote standard letter components. Reference to an attachment is included.

7. signature, which may be optional, depending on company policy.

Unlike letters, informal memos typically need not be signed. The format shown is common, but any neat format that contains all the necessary information is usually acceptable. Most word processors provide letter and memo templates.

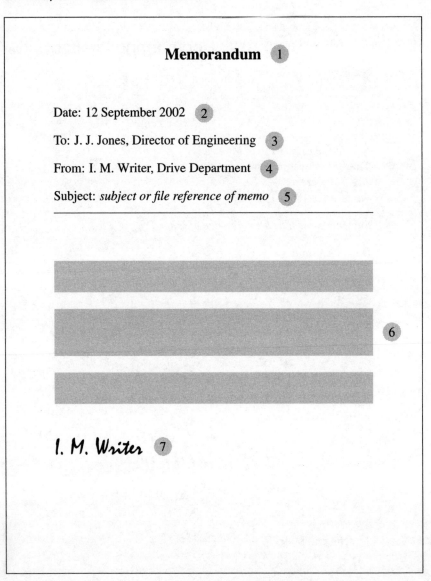

Fig. 6.3. A typical memo. The signature may be optional, depending on company policy.

1.3 Email

Electronic mail sent and received by computers or portable electronic devices is rapidly becoming a standard means of communication. An email message is like a memo that has not yet been printed. Thus email can be informal or formal, although documents that must be signed will have to await the widespread adoption of encrypted digital signatures. The great convenience of electronic messaging raises several issues, discussed briefly in the following paragraphs.

Privacy: Many companies reserve the right to inspect any message sent using their facilities, and some strictly prohibit the use of their facilities for personal purposes. On some systems, the number and size of messages sent and received by an individual is available to any other user of the mail system. Never send an email that you would not want your boss or colleagues to know about or read.

Security: Business email may be sent through the internet, but many large companies maintain their own internal networks for security reasons, and other companies make arrangements for encrypting messages sent through public channels. Never send an unencrypted email containing material that would cause difficulty if disclosed.

Disagreement: Anyone who has witnessed the animosity that is common in internet discussion groups will be aware that debate quickly tends to become hardened or acrimonious from lack of the eye contact, body language, and voice tone that are important parts of face-to-face communication. If you need to resolve a difference of opinion with someone, speak to them personally or use the telephone, in preference to debating by email. The exchange of memos can also be potentially acrimonious, but email can be sent much more quickly, and can lead to interpersonal difficulties that should not exist in a professional environment. Therefore, if you must debate with someone via email, pay special attention to being polite, wait a day if possible before returning messages, and carefully reread your replies before sending them.

1.4 Specification documents

In order to obtain an engineering product or service, a client or purchaser usually writes a specification document containing the criteria that must be satisfied by the product or service. The specification document is used by the engineers who design, build, or otherwise provide the product. Thus a specification contains the criteria that the desired product or service must satisfy, and may form part of a contract between the client and provider. The client may be an external supplier or another department within the same organization.

A specification document is basically a list of criteria or tests that determine the characteristics required of a desired product, component, process, or system. Thus, for example, an airline company might produce a set of specifications for an airplane that they wish to purchase, listing the required payload, range, speed, operating environment, internal fixtures, delivery schedule, payment constraints, and other factors. Manufacturers would then bid on the proposed contract, adapting their existing airplane types to meet the desired specifications, and would likely have to go through a similar process of writing specifications to obtain the engines, electronics, landing gear, hydraulic systems, and the many other components required. The process is repeated at increasing levels of detail. Thus, specification documents may be used from the most general level to describe complete airplanes, down to the detailed level to describe the size and strength of individual bolts, for example.

A specification document generally contains the following:

1. Introduction and scope: the general purpose of the product or service required, with an overview of the range of application.

2. List of requirements: Each requirement that must be satisfied is listed, possibly with the procedures to be used to test whether the requirement is satisfied.

1.5 Bids and proposals

Bids and proposals are offers from engineers to provide services. A *bid* is an offer to provide specified services; a *proposal* typically suggests a means of meeting a need that has not been specified precisely, and offers to provide the required service.

Thus, a potential client may send a "request for quotation" to eligible bidders, including specifications for the job to be done. The response to the client is a bid document, which states how the engineering work will comply with the specifications and any exceptions that are proposed, together with a price at which the company is willing to do the work. Other information may be required by the client or offered by the engineer, such as the qualifications of the individuals who will perform the work. Government agencies tend to have strict rules for the format and timing of bid documents.

When required engineering services are not precisely defined, and there are several possible solutions, then a "request for proposal" is issued. The engineer is then free to outline a solution and to offer to provide it. Since a proposal is in part intended to convince a prospective client of the competence of the engineering provider, the proposal document is produced using a permanent binding, in a formal but attractive style, and is accompanied by a letter of transmittal as discussed for engineering reports in Chapter 8.

Bid and proposal documents, when accepted, may form part of a legal contract that binds the client and engineering group together for the project.

1.6 Reports

Reports may be produced in many forms. Usually the production of the document has been commissioned by a superior or client, either to answer specific questions, to investigate a situation in the interest of the client, or simply to provide a record of an event or situation. Therefore, reports may be technical or non-technical, formal or informal, depending on the circumstances.

Formal technical reports may be the official record of a company's conclusions or actions, and may have to withstand the scrutiny of a legal proceeding, for example. Chapter 8 discusses formal technical reports typically required within a company, by a client contracting for engineering services, or from a student returning from an industrial internship or work-term.

In the following, the logical structure of reports is described, together with their physical structure. This section then discusses experimental and laboratory reports, evaluation reports, and progress reports.

Logically, there are three main components in a report:

- *Introduction*—the purpose and background of the work presented in the report. The questions being investigated are posed, together with background information required by the intended readership.
- *Detailed content*—the investigations, results, or analysis satisfying the purpose of the report, in as much detail as the intended readers are expected to require. The material is presented in a clear, logical sequence, using sections and sub-sections as necessary.
- *Conclusions*—the answers to the questions posed in the introduction. In a long report these answers may have appeared in the detailed content and are summarized in this section; in a shorter report the conclusions supported by the analysis are stated. It should be possible to pair each conclusion with a question posed in the introduction.

Many reports are similar to books in structure, and contain the following parts:

- *Front matter*—components such as the title page and table of contents, which introduce and index the document. In the past, the front matter could be composed only after the body of the document was finished, and the page numbers of the body had been determined. Therefore, the front matter had to be given a separate set of page numbers, printed in lower-case roman numerals. This numbering tradition persists, even with the use of modern production by computer.
- *Body*—material divided into numbered sections, beginning with the Introduction section and typically ending with the Conclusions section, with other sections and subsections as necessary to present the material clearly and logically to the target audience.
- *Back matter*—material that supplements the contents of the body but is not essential for understanding the report. The back matter may also contain material that is too large or that would otherwise break up the flow of the body. Typically the back matter contains lists of references, bibliographic material, or appendices. Appendices are normally labelled Appendix A, Appendix B, and so on, rather than numbered.

Report formats vary according to their formality, purpose, and content. However, specific formatting rules may be imposed by the client or by the company producing the report. It is always wise to check for special requirements, and when several documents of the same type are to be produced, to develop a document template in the correct format, thus ensuring uniformity.

The format of a formal engineering report is discussed further in Chapter 8, but some other common report types are briefly described below:

Experimental and laboratory test reports: All engineering students will have to write laboratory reports during their degree programs. The purpose of the laboratory is typically to illustrate or test theoretical concepts studied in class or as a vehicle for learning the test procedure, and the report is written according to the principles of the "scientific method," which requires the objective presentation of the work and the conclusions that follow from it. A typical list of headings is shown in Table 6.1. The main part of the report consists of the numbered sections, beginning with the Introduction section and ending with the Conclusions section.

In business and industry, the purpose of the technical report is rarely to verify newly learned theory; the report is usually intended to determine the following:

- the feasibility of a new process, product, or device to be used, for example, in the design of a new product,
- the results of a standardized test of performance or safety,
- exploratory data, such as soil tests for building foundation design.

A test report generally must contain all of the details of the apparatus, procedure, and test conditions. However, many reports are the result of industry-standard tests, routinely conducted, such as the verification of the properties of concrete or tests of the hardness of metal alloy. In such cases, the report may consist only of an introduction, reference to the industry standards defining the tests, the measured data with its analysis, and resulting conclusions.

Evaluations of products, processes, or feasibility: Product or process evaluations may be required for product improvements, to investigate unexpected performance, or to judge the qualities of an alternative or competing product. In all cases, the introduction must carefully state the criteria by which the evaluation is being made, and the conclusions must compare performance to the stated criteria.

Feasibility reports are similar to evaluations but discuss projected or hypothetical products or processes, rather than those that are available. Prototypes of proposed products are sometimes the subject of feasibility reports, in order to predict the performance of the final product.

Evaluations and feasibility reports have a common structure: the performance of the whole is determined by examining its parts, comparing their performance with respect to stated criteria, and reaching global conclusions by combining the stated criteria by a logical decision method such as is discussed in Section 5 of Chapter 16.

Table 6.1. Components of a typical laboratory report.

Title page	Department name, course name and number, professor name, student name and number or other required identification, report title, date.
Summary	A summary in less than one page of the report contents and conclusions.
Contents	A list of the report section and sub-section titles and numbers, with page numbers.
List of figures	The figure numbers and captions, with page numbers.
List of tables	The table numbers and captions, with page numbers.
1. Introduction	The purpose and circumstances under which the experiment was conducted and the report written. The questions to be answered in the report must be posed.
2. Theory	The theoretical concepts required to understand the experiment and the hypotheses that are to be tested.
3. Equipment	A description of the equipment required, given in sufficient detail to duplicate the setup in equivalent circumstances.
4. Procedure	The methods used to obtain the observations and data recorded in the report.
5. Results	The data recorded with qualitative observations as necessary.
6. Analysis	Computations with graphs or other information required to compare the data to the hypotheses tested.
7. Conclusions	The answers to the questions posed in the Introduction.
References	Author, title, and publication data in a standard format for the documents cited in the report.
Appendices	Large tables, diagrams, computer programs, or other material that is not meant to be read as part of the body of the report.

Progress reports: During an engineering project it may be necessary to send reports to company management or to clients to describe what parts of the work have been completed or are under way. These reports may also be written as an internal record of steps undertaken and accomplished. The primary question to be answered is whether the actual progress corresponds to the budgeted time, financial and physical resources, and human expertise. The content of these reports can form a partial draft of the final report.

2 Further study

1. Check your word processor to see if it has a standard format or template for writing technical letters and memoranda. How do the templates differ from the formats suggested in this chapter?

2. Prepare word-processor templates of (a) a laboratory report for use in your own classes and (b) a formal engineering report that you could use to report on an industrial internship or work-term.

3 References

[1] D. Beer and D. McMurrey, *A Guide to Writing as an Engineer*, John Wiley & Sons, New York, 1997.

[2] R. Blicq and L. Moretto, *Technically-Write!*, Prentice Hall, Scarborough, ON, 1998.

[3] J. N. Borowick, *Technical Communication and its Applications*, Prentice Hall, Upper Saddle River, NJ, 2000.

[4] K. W. Houp, T. E. Pearsall, E. Tebeaux, S. Cody, A. Boyd, and F. Sarris, *Reporting Technical Information*, Allyn and Bacon Canada, Scarborough, ON, 1999.

[5] D. Jones, *The Technical Communicator's Handbook*, Allyn and Bacon, Toronto, 2000.

[6] D. G. Riordan and S. E. Pauley, *Technical Report Writing Today*, Houghton Mifflin Company, New York, 1999.

[7] B. R. Sims, *Technical Writing for Readers and Writers*, Houghton Mifflin Company, New York, 1998.

[8] J. S. VanAlstyne, *Professional and Technical Writing Strategies*, Prentice Hall, Upper Saddle River, NJ, 1999.

[9] K. R. Woolever, *Writing for the Technical Professions*, Addison-Wesley Educational Publishers, Inc., New York, 2002.

Chapter 7

Technical writing basics

When graduate engineers are surveyed about key skills for career success, communication skills are inevitably mentioned. You must be a competent writer. You must be able to use and learn from a dictionary, a thesaurus (Figure 7.1), and a technical writing reference kept near your desk, or their software equivalents. You must also be proficient with software tools for word processing, for creating diagrams, and for managing bibliographies. Like an athlete, you should exercise your skills regularly and continually attempt to improve them.

In this chapter, you will learn

- hints for improving your writing style,
- some of the most common writing errors and how to avoid them,
- a review of basic English punctuation,
- hints and a summary on parts of speech,
- hints for the use of Greek letters in technical writing.

The questions at the end of this chapter include a self-test for you to gauge your writing ability.

516 Intelligibility

N. *intelligibility*, knowability, cognizability; explicability, teachability, penetrability; apprehensibility, comprehensibility, adaptation to the understanding; readability, legibility, decipherability; clearness, clarity, coherence, limpidity, lucidity 567 n. *perspicuity*; precision, unambiguity 473 n. *certainty*; simplicity, straightforwardness, plain speaking, plain speech, downright utterance; plain words, plain English, mother tongue; simple eloquence, unadorned style 573 n. *plainness*; easiness, paraphrase, simplification 701 n. *facility*; amplification, popularization, haute vulgarisation 520 n. *interpretation*.

Adj. *intelligible*, understandable, penetrable, realizable, comprehensible, apprehensible; coherent 502 adj. *sane*; distinguishable, audible, recognizable, unmistakable; discoverable, cognizable, knowable

Fig. 7.1. Part of an entry from a thesaurus [1] is shown. An English dictionary, a technical dictionary, a thesaurus, and their electronic equivalents are essential tools for producing correct technical documents.

1 The importance of clarity

Engineers often work in teams, and effective teamwork depends on effective communication. Moreover, engineering work normally must be recorded for future reference. Decisions affecting worker or client health, property, or public welfare must be carefully recorded. In the event that your company has to demonstrate that it has followed standard professional practice, your documents might be scrutinized in a court of law. For these and many other reasons, you must write unambiguously and clearly.

The remainder of this chapter contains some points to keep in mind as you analyse the clarity and correctness of your writing. However, you must also know how to describe and interpret inexact quantitative data, using the methods of Part III, and to design graphs, drawings, and other artwork as described in Chapter 9, since a picture can explain complex relationships that are difficult to express verbally. The logic-circuit diagram in Figure 7.2, for example, shows at a glance how a particular combination of logic elements is constructed. Similarly, a surveyor's plan shows the location and shape of a piece of land accurately and concisely, but a lawyer's written description of the same piece of land might require several pages of legal jargon. Although a full discussion of engineering drawing is beyond the scope of this textbook, you should develop your ability to sketch and to make technical drawings using software and other tools.

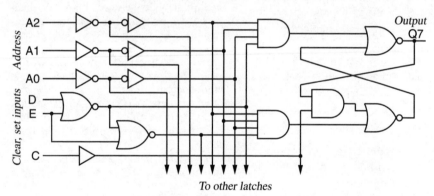

Fig. 7.2. A diagram shows the connections of a binary logic circuit more clearly than could be done using words. This circuit is called a general-purpose latch. Inputs A0–A2 are address lines, and the inputs C, D, E are used for setting and clearing the output Q7.

2 Hints for improving your writing style

Practise your skills consciously whenever you have to write, consulting a dictionary, thesaurus, university-level grammar handbook, or electronic equivalents of these to hone your skills. Those who suggest that they have no need for aids or for improving their writing skills are like athletes who suggest that they have no need to train. Keep handy an English reference such as [2] or [3]. If your writing includes significant amounts of mathematical analysis or formulas, refer to [4].

Clarity and brevity: In technical writing you will be mainly concerned with communicating information, so your style should be clear and concise. Say what you mean, as briefly and clearly as possible. However, while short sentences are concise, a report containing only short sentences is tedious to read; vary the sentence length to avoid monotony.

Correct terms: Define any specialized terms, either in the text or in a glossary, that your intended reader may not understand. Avoid slang, jargon, or compound words that do not appear in a dictionary. Consult a technical dictionary, as required, to verify your usage. General technical dictionaries such as [5] may suffice, but you may also have to consult a specialized dictionary, such as for construction [6] or electronics [7]. In many fields there are also on-line reference dictionaries and collections of them, for example, reference [8].

Specific terms: Use specific, concrete words rather than generalities. For example, in describing a manufacturing operation, the statement: "A hole was made in the workpiece" may be true, but *made* is a weak, general word. It is more informative to say: "A 12 mm hole was drilled through the workpiece." Similarly, "The machinery cannot be moved because the floor is bad" may be true, but be specific: "The wooden floor has several rotten planks and is unlikely to support the machinery adequately." Avoid generalities and overuse of terms such as *to be, to have, to make,* and *to use.*

Spelling and punctuation: Poor spelling and punctuation will destroy the confidence of the reader in your work. If you are in doubt about the meaning or spelling of a word, consult a dictionary. If you have difficulty choosing precisely the right word, consult a thesaurus. Your computer may have a thesaurus or spell checker; however, these tools cannot identify correctly spelled words substituted for others, such as in the pairs

- form and from,
- there and their,
- dough and doe,
- ferry and fairy,

or the triples

- your, you're, and yore,
- for, four, and fore.

You must proofread what you have written. See Section 3 for a basic summary of punctuation.

Tense: Inconsistency of tense and excessive use of the passive voice are common report-writing errors. The tense of a verb indicates the time of the action or state. For most practical purposes, there are 6 tenses, as illustrated in Table 7.1, although purists double this number to 12 by considering certain auxiliary forms [2]. Since

72   Chapter 7   Technical writing basics

Table 7.1.   Chart of tenses for the verb *to see* in both active and passive voice. All tenses are technically correct, but the less desirable wording has been shaded.

Tense	Active voice	Passive voice
Present	I (We) see. You see. He (She, It) sees. They see.	I am (We are) seen. You are seen. He (She, It) is seen. They are seen.
Past	I (We) saw. You saw. He (She, It) saw. They saw.	I was (We were) seen. You were seen. He (She, It) was seen. They were seen.
Future	I (We) will see. You will see. He (She, It) will see. They will see.	I (We) will be seen. You will be seen. He (She, It) will be seen. They will be seen.
Present perfect	I (We) have seen. You have seen. He (She, It) has seen. They have seen.	I (We) will be seen. You have been seen. He (She, It) has been seen. They have been seen.
Past perfect	I (We, You) had seen. He (She, It, They) had seen.	I (We, You) had been seen. He (She, It, They) had been seen.
Future perfect	He (She, It, They) will have seen.	He (She, It, They) will have been seen.

most technical reports are written after measurements or other work are completed, the present or past tenses are most appropriate. Write in the future tense only when you are discussing an event that will occur in the future, such as in a section discussing suggestions for future work. Do not use the future tenses to refer to something appearing further along in the report.

Voice: There are two voices: active and passive, illustrated in Table 7.1. The active voice is the more emphatic and interesting, as seen by comparing the following two examples:

Active: "The machine operator repaired the damaged workpiece."

Passive: "The damaged workpiece was repaired."

Use the passive voice if the doer of the action is irrelevant to the discussion and the active voice to produce shorter, clearer sentences. For example, use the active voice

- to bring a main idea to the front of a sentence: "Figure 1 shows ...,"
- to emphasize an identity: "Onitoba Manufacturing provided the specially designed clamps ..."

Person: The three choices of person, first (I, we), second (you), and third (he, she, it, they), are illustrated in Table 7.1.

Use the first person

- to take responsibility, for example, "We recommend ...,"
- to place yourself in an active role, for example, "By the third interview, we observed that the attitude of staff members had changed."

The second person is often implicit in instructions; for example, the subject *you* is implied in, "Close the feed valve when the liquid level reaches the red line." The word *you* should mean the reader, not people in general.

Most technical writing should be objective and independent of the observer. Therefore, the third person is sometimes most appropriate, particularly in descriptive parts of formal technical reports. However, the third person leads to passive, complicated sentence structure; do not overuse it.

Questions of tense and person may be arbitrarily resolved if your employer (or university course) requires you to write in the third person; if so, then follow the rule but include plenty of active sentences. The authors discussed voice and person while writing this textbook. We decided that active sentences are more interesting, so you will find that the first and second person are used frequently in this edition.

3 Punctuation: A basic summary

Punctuation is a code to tell you where stops, pauses, emphasis, and changes in intonation occur. The 13 basic punctuation symbols are explained briefly below.

Period: The period (.) is a "full stop," which ends a sentence. It is also used after most abbreviations, for example, *Mr., Ms., Dr., Prof.*

Comma: The comma (,) indicates a slight pause to separate ideas, such as to separate two independent clauses joined by a conjunction; for example, "School was hard, but life was full." The comma is also used as follows:

- to give a pause after transitional words such as *however, nevertheless, moreover, therefore;*
- to separate a long introductory phrase from a principal clause; and
- to separate words, phrases, or clauses in a list of three or more like items.

Commas are often used in pairs, like parentheses, to enclose words, phrases, or clauses that add non-essential information. The meaning of the sentence must not change if the commas and enclosed material are deleted. As a general rule, a single comma should not separate the subject from the verb or the verb from the object.

Semicolon: The semicolon (;) indicates a longer pause than a comma but shorter than a period. It is used between two main clauses that express a continued thought but are not joined by a conjunction; for example, "Intellectuals discuss ideas; gossips discuss people." It is also used between two clauses that are joined by a conjunctive adverb; for example, "It rained; fortunately, the wedding was indoors." The semicolon is also used in place of a comma in a list if the phrases or clauses in the list contain commas.

Colon: The colon (:) usually introduces a list of like items, as a substitute for *namely*, or *for example*. It should not be used unnecessarily; for example, after a verb or after *of*. The colon is also used to introduce a formal statement, quotation, or question; for example, "The valve was marked: Do not open!"

Hyphen: The hyphen (-) joins compound adjectives; for example, *natural-gas producer, two-month vacation*. A long word that does not fit on a line is hyphenated between syllables and completed on the next line. Hyphens are also used in compound nouns, for example, *passer-by, dry-cleaning*. There are no clear rules about the construction of compound nouns; consult a reference when in doubt.

Dash: The dash (—) shows a sudden interruption in thought. For example, "It stopped raining—I think!" The dash is common in dialogue but is rare in formal writing.

Question mark: The question mark (?) is a full stop at the end of a direct question; for example, "May I help you?" However, an indirect question ends with a period; for example, "He asked if he could help you."

Exclamation point: The exclamation point (!) is the full stop at the end of a sentence that expresses surprise or strong feeling; for example, "We won! I can't believe it!" The exclamation point is used rarely in formal writing.

Apostrophe: The apostrophe (') has two uses. The first is to indicate possession. It is used at the end of a noun, followed by the letter *s*; for example, "He has writer's cramp." The apostrophe also replaces the deleted letters in abbreviations; for example, *doesn't; it's*. Note that *it's* always means *it is* and does not mean *belonging to it*.

Quotation marks: Quotation marks (" ") enclose direct quotes and, when the style requires it, article titles in a list of references.

Brackets: Brackets ([ ]) enclose material added editorially to clarify meaning; for example, "The boss [Mr. George] was described as being sick."

Parentheses: Parentheses "( )" enclose ideas of secondary or alternate importance. An idea placed in brackets or parentheses must not be essential; the meaning should not change if the parentheses and the enclosed material are deleted.

Ellipses: Ellipsis points (...) are three spaced points used in place of omitted words.

4 The parts of speech: A basic summary

Proper sentence structure is the key to clear writing. Sentences are created from eight common parts of speech: nouns, pronouns, verbs, adjectives, adverbs, prepositions, conjunctions, and interjections. These terms signify not what a word is, but how it is used, since the same word may be used in several ways. To refresh your memory, the parts of speech are defined below:

Noun: A noun names a thing or quality; for example, *a box, beauty*. A proper noun names a specific person or place and has an initial capital letter; for example, *Fred, Waterloo*.

Pronoun: A pronoun takes the place of a noun; for example, *it, her, him, anyone, something*. Each pronoun refers to a noun, which is called the antecedent of the pronoun. The pronoun and its antecedent must agree, and the antecedent must be clear and unambiguous; for example, "The box broke; burn it," and "Fred called; please call him."

Verb: A verb is often called an action word, since it indicates action, although it can also indicate a state of being; for example, *run, jump, make, be, exist*.

Adjective: An adjective modifies nouns; for example, a *big* box; a *tall* building, a *small* raft.

Adverb: An adverb modifies verbs, adjectives, or other adverbs; for example, *rapidly, softly*.

Preposition: A preposition is also a linking word, for example, *in, on, of, under, over, to, across*. A preposition is used to show the relationship of a noun or pronoun to some other word in the sentence, for example, *depth of water, voltage across the terminals, on a shelf*.

Conjunction: Conjunctions are linking words, for example, *and, but,* and *or;* they join words, such as *bat and ball*. They may also join phrases, for example, *date of arrival and length of stay,* or clauses.

Interjection: Interjections express emotion, for example, *Oh!, Ouch!, Wow!,* and are rarely used in technical writing.

5 Avoid these writing errors

This section reviews some basic errors that you should avoid. Such a brief discussion cannot cover every topic; the handbooks in the references should be consulted for further explanations and additional topics.

Sentence fragments: The sentence is the basic building block of written communication and contains a complete thought. However, in the fury of composition, sometimes writers leave fragments in their work that have partial meaning but are not complete sentences. For example, "The new, advanced gear-box. Delivering over 200 HP!" In spoken dialogue, the speaker's tone of voice and emphasis help us to fill in any missing information. However, in a formal written report, sentences must be complete.

Dangling modifiers: The dangling modifier is a phrase, usually at the start of a sentence, that fails to identify what it modifies. A typical example of this common error is, "Although only 16 years old, the policeman treated him like an adult." This sentence implies that the policeman was only 16 years old, whereas the writer probably meant that the subject of the policeman's attention was 16 years old. The reader expects the phrase to modify the subject of the sentence, in this case, the policeman. Avoid this common error by asking yourself, whenever you write a modifying phase, "What does it modify?"

Comma splices and run-on sentences: Sometimes writers splice two principal clauses into a single sentence, using a comma. For example, "The voltage limit was exceeded, the capacitor failed." Correct this error by splitting the sentence or inserting a semicolon or conjunction. For example, "The voltage limit was exceeded. The capacitor failed," or "The voltage limit was exceeded; the capacitor failed," or "The voltage limit was exceeded, and the capacitor failed." A run-on sentence is an even more basic error; it is a comma splice in which the writer left out the comma.

Superfluous commas: Sometimes writers add superfluous (excess) commas. For example, "The data were read but, not organized in charts. Perhaps, the data were written down incorrectly." Extra commas confuse the reader. Although every comma indicates a pause, not every pause requires a comma. For example, "The data were read but not organized in charts. Perhaps the data were written down incorrectly." It is worth spending the time to review the reasons for using the comma, defined in detail in Section 3.

Subject-verb disagreement: The appearance of an incorrect verb may be a sign of an excessively complicated sentence, as well as a basic error of grammar. The most common error is the incorrect plural verb form illustrated by "The insertion of zener diodes stabilize the circuit."

Adverb and adjective confusion: Adverbs and adjectives are both modifiers. Do not confuse them. Adverbs frequently end in *ly,* for example, *rapidly, softly,* and *quickly.* Adjectives modify only nouns, but adverbs may modify verbs, adjectives, or other adverbs. It is correct to say "a really big box" but not correct to say "a real big box," since *really* is an adverb, but *real* is an adjective and cannot modify the adjective *big.* If you are not sure whether a word is an adverb or an adjective, consult your dictionary.

6 The Greek alphabet in technical writing

Technical writing often requires mathematical notation that may contain Greek letters. Use them sparingly, particularly if the report is intended for readers without mathematical expertise. Consult [4] for detailed advice about writing mathematics. The Greek alphabet is shown in Table 7.2.

Table 7.2. The Greek alphabet, showing English equivalents. The meaning attached to each symbol varies with the context.

Greek letter	Greek name	English equivalent	Greek letter	Greek name	English equivalent
A α	Alpha	a	N ν	Nu	n
B β	Beta	b	Ξ ξ	Xi	x
Γ γ	Gamma	g	O o	Omicron	ŏ
Δ δ	Delta	d	Π π	Pi	p
E ϵ	Epsilon	ĕ	P ρ	Rho	r
Z ζ	Zeta	z	Σ σ	Sigma	s
H η	Eta	ē	T τ	Tau	t
Θ θ	Theta	th	Υ υ	Upsilon	u
I ι	Iota	i	Φ ϕ	Phi	ph
K κ	Kappa	k	X χ	Chi	ch
Λ λ	Lambda	l	Ψ ψ	Psi	ps
M μ	Mu	m	Ω ω	Omega	ō

7 Further study

1. The simple test below has 30 questions in total in Parts A, B, and C. Give yourself about 20 minutes to complete it. Then check the answers at the end of the chapter.

Scoring: 27–30: Excellent;   24–26: Good;   20–23: Fair;   < 20: Poor.

A. Vocabulary: In the following sentences, select the appropriate word:

(a) The kids are (all right, alright).

(b) In his fall, he narrowly (averted, escaped) death.

(c) The committee voted to (censure, censor) him for his poor behaviour.

(d) The engineering faculty is (composed, comprised) of six departments.

(e) The police siren emitted a (continuous, continual) scream.

(f) I am indifferent to her wishes; in fact, I (could, could not) care less.

(g) Her hint (inferred, implied) that she had a gift for him.

(h) We (should of, should have) removed the specimen before it broke.

(i) This machine is more reliable, since it has (fewer, less) parts to fail.

(j) The conference is (liable, likely) to be held at the university.

(k) He re-wrote the sub-routine (his self, himself).

(l) He asked me to (loan, lend) him my lawnmower.

(m) She bought a small (momento, momentum, memento) as a souvenir.

(n) It was so hot that I (couldn't hardly, could scarcely) breathe.

(o) Asbestos is (non-flammable, inflammable).

B. Grammar and usage: In each of the following sentences, state whether the grammar and usage are correct or incorrect.

(a) Americans speak different.

(b) She hit me hard on the arm.

(c) The technician took a real big wrench out of the tool-box.

(d) My sister, always bugging me.

(e) The four students were quarrelling between themselves.

(f) What about those Bluejays!

(g) Driving down the road, the poles went past us quickly.

(h) Check you're spelling thoroughly.

(i) Whom do you recommend?

(j) Torquing is when you tighten a nut with a wrench.

C. Punctuation: In each of the following sentences, state whether the punctuation is correct or incorrect.

(a) She said, "I am sure the fuse is broken; it happens frequently."

(b) He shouted, "The transmission lost it's oil!"

(c) They all drove over here in John's and Jean's cars.

(d) The crowd was mad, it strung him up.

(e) He bought ten dollars' worth of candy for the children's party.

2. Correct or improve each of the following statements:

(a) Never use no double negatives.

(b) In his defence, he respectively told the magistrate that the auto was stationery at the time of the accident.

(c) A verb have to agree with its subject.

(d) Your represented by your writing.

(e) Clichés should be as scarce as hen's teeth.

(f) Proof-reed everything, especially watch you're spelling.

(g) Eschew obfuscation in whatever manifestation.

(h) Make sure colloquial language ain't used.

(i) When dangling, subjects are not properly modified.

(j) Throughly check you're spelling and punctiliousness.

3. The following sentences contain word usage or clichés that should be avoided. In each case, suggest a simpler or alternative way to state the idea.

(a) The machine is not functioning at this point in time.

(b) The project is inoperative for the time being; it's on the back burner.

(c) In the stock market, fear and greed dominate; this is the name of the game.

(d) You know, basically, I'll have the chocolate ice cream.

(e) The three alternatives caused him a dilemma. By and large, the bottom line is consumer driven.

4. Find a grammatical error or punctuation error in the body of this textbook and write a letter of reprimand to the authors. Be polite but firm.

Answers, Problem 1:

A. Vocabulary: (a) all right (b) escaped (c) censure (d) composed (e) continuous (f) could not (g) implied (h) should have (i) fewer (j) likely (k) himself (l) lend (m) memento (n) could scarcely (o) non-flammable.

B. Grammar and usage: (a) Americans speak differently. (b) Correct. (c) ... really big wrench ... (d) My sister is always bugging me. (e) The four students were quarrelling among themselves. (f) What do you think about those Bluejays? (g) Driving down the road, we saw the poles go past us quickly. (h) Check your spelling thoroughly. (i) Correct. (j) Torquing is the act of tightening a nut ...

C. Punctuation: (a) Correct. (b) ... lost its oil ... (c) Correct. (d) The crowd was mad; it strung him up. (e) Correct.

8 References

[1] R. A. Dutch, *Roget's Thesaurus*, Longmans, London, 1963.

[2] D. Hacker, *A Canadian Writer's Reference*, Nelson Thomson Learning, Scarborough, ON, 2001.

[3] K. R. Woolever, *Writing for the Technical Professions*, Addison-Wesley Educational Publishers, Inc., New York, 2002.

[4] N. J. Higham, *Handbook of Writing for the Mathematical Sciences*, Society for Industrial and Applied Mathematics, Philadelphia, 2nd edition, 1998.

[5] R. Ernst, *Comprehensive Dictionary of Engineering and Technology: With Extensive Treatment of the Most Modern Techniques and Processes*, Cambridge University Press, Cambridge, 1985.

[6] L. F. Webster, *The Wiley Dictionary of Civil Engineering and Construction*, John Wiley & Sons, New York, 1997.

[7] Institute of Electrical and Electronics Engineers, *IEEE Standard Dictionary of Electrical and Electronics Terms*, Institute of Electrical and Electronics Engineers, New York, 1977.

[8] J. Kirtland, *Academic Press Dictionary of Science and Technology*, Harcourt, Inc., New York, 1999, <http://www.harcourt.com/dictionary/> (February 15, 2002).

Chapter

Formal technical reports

The formal technical report is the definitive presentation of the results of engineering work. The designation "formal" refers primarily to the logical structure and writing style, rather than to the purpose or application. A report might record the investigation of an accident, the analysis of a failure, or the feasibility of a design. In fact, the subject matter of formal reports is as broad as the field of engineering.

Above all other considerations, a formal report must be correct, clear, objective, and complete. It should be carefully and artistically produced, with professional-quality layout and graphics (Figure 8.1).

In this chapter, you will learn

- the purpose and format of the logical and physical report components,
- a strategy for writing reports,
- a list of items to check when completing a report.

Fig. 8.1. A well-produced technical report makes a subtle statement about the quality of its content. For details on cover components, see page 82.

The great majority of report documents can be produced using the suggestions given in this chapter. However, rules occasionally have to be varied or broken according the objectives of the report and when demanded by the need for clarity. With your word processor you can and should create templates of the main kinds of reports that you may have to produce in the future.

This chapter includes only the most important considerations for report writing. Whole books have been written on the subject and on technical communication in general; see [1], [2], and [3], for example, and for advice on writing material containing mathematics, see [4]. Consult your library for others.

1 Components of a formal report

Readers will most easily understand material presented in a logical and familiar format. Therefore, the rules for defining report components are based both on logic and on tradition or familiarity. Table 8.1 lists generic headings for technical reports; however, the section headings in the body should be changed to be more descriptive of the subject at hand.

Your company may have specific rules about the format of formal reports, and the headings in Table 8.1 may have to be modified to conform. You may be given a report template that specifies formatting detail such as heading and text font, document margins, and other details.

In many ways a report is similar to a book, although reports are not normally divided into parts and chapters as this book is, but into sections and sub-sections as are the book chapters. The front matter of this book can be used as a model for a report, although a report typically does not have a preface.

Technical reports can be divided physically into three parts: the front matter, the body, and the back matter. The front matter and back matter contain material that supplements and supports the body of the report. These parts were discussed briefly in Section 1.6 of Chapter 6, but formal reports have more specific requirements, as described in the next sections.

1.1 The front matter

The number of pages and the details of the front matter can be known only when the rest of the report is complete. Traditionally, to avoid renumbering all of the document pages, front matter pages were numbered independently and distinctly, using lower-case roman numerals (i, ii, iii, ...). The use of computers has removed the technical need for separate page numbers, but tradition is strong and we continue to number the front matter with roman numerals; a report that does not do so is considered to be incorrect.

The title page is always page i, but it is artistically unsuitable for the number to appear there, so usually the first visible page number is on page ii or later.

The components of the front matter are listed in their usual order in the following paragraphs, although minor placement variations are possible.

Cover: Your company may have standard binding and cover requirements such as illustrated in Figure 8.2. The title of the report must be clearly visible on the external cover or binder; the reader should not have to open the report to see the title. In the absence of a special design, a clear plastic front cover that allows the title page to be read without opening the report is usually acceptable.

The standard components for the cover are:

Table 8.1. Components of a formal report. The body contains numbered sections, usually more than the four shown, with sub-sections as necessary. It is preceded by the front matter, which is page-numbered with roman numerals, and followed by the back matter. Items marked with an asterisk (*) are included only when necessary.

Front matter		Front cover
		Title page
	*	Letter of transmittal
	*	Preface
	*	Acknowledgements
		Summary (or Executive Summary)
	*	Abstract
	*	Key words
		Contents
	*	List of figures
	*	List of tables
	*	List of symbols
Body	1.	Introduction
	2.	Analysis
	3.	Conclusions
	* 4.	Recommendations
Back matter		References
	*	Bibliography
	*	Appendices

- title,
- author's name,
- date of publication,
- author's affiliation: company and department name.

Other items, such as library cataloguing information, the name of the recipient, or the author's student identification number and department, may be included on the cover as necessary.

Title page: The title page is the ultimate condensation of the report and must contain the following:

- the report title,
- author's name and address,

ABC Engineering Limited

**Design modifications
to the
model 894 hydrogenator
for optimized pressure**
(Contract 200-2002 final report)

Prepared for:
E. Smith, Manager
Pillz Pharmaceuticals Ltd.

22 September 2002

Chris Cross, P.Eng.
Contract Design Department

123 Fourth Street
Anywhere ON N2L 3G1
(999) 123-4567,   ABCEngineeringLimited.com

Fig. 8.2. A typical report cover defined by a company. The key components are shown: title, "Prepared for" information, date, author, and author's affiliation.

- date of report completion,
- the name and affiliation of the person for whom the report was prepared.

A title should not be too short for the intended reader to understand the report purpose. For example, consider a report investigating the cause of fires in a particular model of electrical control panels. Here are some example titles:

- Control Panel Fires
- The Cause of Six Fires in the Model 24 Control Panel
- The Cause of Six Fires in Paneltronix Model 24 Control Panels Installed in Municipal Pumping Stations

The first title is much too vague for any circumstance. The second is adequate for a reader who knows what a Model 24 panel is but still too vague for a more general readership, for which the third probably is sufficiently clear and complete.

Choose precise, concrete words for the title, and avoid superfluous words such as *Report on ...*, or *Study of ...* Do not merely name an object or machine; say something about it. For example: *Number Six Forging Press* identifies the machine but is vague. *Motor Replacement on the Number Six Forging Press* identifies the modification and the machine and might be much better, depending on the intended reader. A report title page may show the same basic information as the report cover, as shown in Figure 8.2, but must be adapted for specific contexts.

Letter of transmittal: Normally a letter accompanies a formal report when it is delivered to the recipient, as a courteous, formal reply to the person who requested it. When the letter is meant to be seen by all the readers, a copy is bound into each report, usually immediately after the title page. The letter has a normal business format and is addressed to the official client or recipient of the report. It must

- refer to the report by title,
- identify the circumstances under which the report was prepared, for example, "in response to your letter of ...," "in fulfillment of contract number ...dated ...," or "for academic credit following my work-term at XYZ Company,"
- identify the report subject and describe the purpose of the report.

In addition, the letter should

- direct attention to parts of the report that should receive particular attention,
- mention recommendations that require action,
- state how the recipient can ask for further information,
- acknowledge assistance received, unless there is explicit mention of assistance in the report,
- direct attention to the engineer's invoice for services, if it is attached.

Preface: A Preface section will normally appear when the letter of transmittal is not bound with the report and you want to give a message to the reader, such as advice for knowledgeable readers who need not consult all report sections. The acknowledgement of assistance may also be given in the preface or in a separate section as described below.

Acknowledgements: Ethical behaviour requires an author to acknowledge any help given in producing the report, in obtaining data or analysing it, for the loan of equipment, for permission to use copyrighted material in the report, or for other assistance. A separate Acknowledgements section is appropriate if there is no preface or bound transmittal letter or if the acknowledgements are extensive or formal.

Summary (or Executive Summary): A technical report is not a novel in which the conclusion is cleverly concealed for 300 pages until the final chapter. The Summary section serves to outline the complete report in advance.

The summary is a complete, independent précis of the report, and it must summarize the introduction, body, conclusions, and recommendations in a few paragraphs, usually less than a full page. For many readers, the summary is the most important section. It is usually written last.

Avoid describing the report in the summary; do not say, "This report describes the modifications to ... ;" say what the modifications are, directly. Similarly, the statement "Recommendations are given" is vague; say what they are.

The title *Executive Summary* is reserved for a summary that is written in non-technical language for persons—often company executives—who are not the primary readers of the report but who typically oversee or finance the project being described. An executive summary usually includes more background description than a normal summary and therefore may be longer, typically up to a few pages.

Abstract: Abstracts are extremely concise summaries, typically limited to 50 to 200 words, suitable for storage in library indexes or other information-retrieval systems. A potential reader may search a library index for material on the topic of the report and should be able to read the abstract to determine the gist of the report contents before obtaining the complete document. Abstracts are required more commonly in published research reports and articles than in privately commissioned reports.

Key words: If the report is to be indexed in an information-retrieval system, a list of descriptive words may be required so that a potential reader can identify documents of potential interest by searching the key word data. A list of key words for this book might contain *professional engineering, introduction to engineering, technical documents, technical measurements, engineering practice,* for example. These are words or phrases that define the subject area of the report. They appear in published technical reports but are more common in technical research articles.

Table of contents: A table of contents is absolutely essential, even in a brief report. The numbered report section and sub-section headings must be listed, with their page numbers. Usually the preface, summary, and lists of figures and tables appear in the table of contents, with their roman numeral page numbers.

List of figures, tables, symbols: In all but small reports, the figures should be listed in the front matter to enable the reader to find them easily. The figure number, the caption or an abbreviated caption, and the page number are given for each figure. A separate list of tables serves a similar purpose, but a report containing only a few figures and tables might combine them into one list.

A list of symbols serves a different purpose: it identifies and defines acronyms, symbols used in formulas, or other special notation. Such a list should be included if symbols have been used with a meaning that the intended readers are not certain to know. The page of first use of each symbol should also be included.

Some reports have a Glossary section in the front matter, containing word definitions as well as symbol definitions.

1.2 The report body

The body contains the essential report material, beginning with an introduction and ending with conclusions and often recommendations. These logical components are organized into numbered, titled sections. Some companies require the Conclusions and Recommendations sections to appear directly after the Introduction section, for easy location and to convey the essential message early, leaving the details for later. The body pages are numbered using Arabic numerals, and the first page of the introduction is page 1.

In the body, the reader is led along a logical path through the sections and sub-sections that support the conclusions and recommendations. Headings in the body must be customized for the topic. The headings depend on whether the report describes a laboratory test, a product design, a field test, a failure analysis, a production efficiency problem, or other topic.

The usual numbering convention for report sections is illustrated below. The title of the introduction is usually Introduction, but the other headings shown must be changed to suit the subject:

1. Introduction
    1.1 Purpose of the report
    1.2 Background information
2. Second section
    2.1 First sub-section of Section 2
    2.1.1 Third-level heading
    2.2 Second sub-section of Section 2
    ⋮

Sub-headings beyond the third level should not be used. The section and subsection titles should be displayed in a distinctive font.

Introduction: The introduction defines the purpose, scope, and methods of the investigation. It must also include sufficient background or history for the intended reader to understand its context. The introduction contains answers to the following:

- What are the purpose and scope of the report; that is, what questions are investigated?

- Why is the report being written; that is, what is the motivation for the work?
- Who or what category of persons is intended to read the report?
- How was the work performed? The process by which the work was done and the report written is outlined.

Sufficient background must be included for the intended readers to understand the body, and the relationship of the report to existing documents cited as references must be described.

The introduction should conclude with a brief outline of the rest of the report, listing what is contained in each of the succeeding sections and the appendices if any.

Internal sections: The subject matter determines the number of sections and their titles. The first of the following alternative ways of ordering sections is preferred when possible:

- in sequence from most important to least important,
- in problem, method, solution sequence,
- in cause-and-effect sequence,
- in chronological (time) sequence,
- in spatial (or location) pattern,
- by classification: group ideas and objects into similar classes,
- by partition: separate ideas and objects into component parts,
- by comparison: show similarities between ideas and objects,
- by contrast: show differences between ideas and objects,
- in order from general to specific.

A large report will not be read completely in one interrupted session; sections of it will be read independently at different times. Include an introductory paragraph or at least an introductory sentence at the beginning of each section, introducing its purpose and outlining its main results.

Diagrams and other artwork: Diagrams, photographs, charts, and tables are often essential, because complex structures and relationships are explained much more clearly by artwork than by written description. Illustrations are discussed in more detail in Chapter 9, but two rules are worth repeating here: each illustration, with an appropriate number and caption, should be inserted in the document just after its first mention in the text; and every illustration must be mentioned in the text.

Citations: The words *citation* and *reference* are sometimes used as synonyms, but usually in the context of document production, a citation is a mention in the report text of an item in the list of references. Each reference in the list, which is in the back matter, provides the title, authorship, and other details of a document that support or supplement the report. The purpose of a citation is

- to place the report in the context of existing documents;
- to mention background material that the reader is expected to know, or at least to know about, in order to read and understand the report;
- to give credit for material that has been quoted: quotations from other documents are permissible if they are clearly distinguished from normal text, and if credit is given to their authors by proper citation.
- to give credit for material that has been rewritten or paraphrased: you are ethically required to acknowledge your sources.
- to mention related or similar material that is not included in the report;
- to allow a conclusion that depends, in part, on work contained in the reference documents;
- to add authority to a conclusion that is confirmed by work in an independent document.

As discussed in more detail in Section 1.3, each reference item has a label, such as "[3]" or "[Jones, 1999]." The label appears in a citation either

- as a noun, for example, "see [3], page 246," or "reference [3] concludes that ..."; or
- as if it were parenthetic, for example, "the symbol % simply means the number 0.01 when used strictly [3]" or "In 1998, Smith and Jones [3] concluded that ..."

Every cited document must be included in the References section (see Section 1.3).

Avoiding plagiarism: As disussed in Section 4 of Chapter 3, plagiarism is the act of presenting the words or work of others as your own. Plagiarism is highly unethical. By using proper citations, you can avoid plagiarism while including brief quotations from other documents in your report, and you can refer to the analyses and conclusions of others. The following guidelines will help you to avoid plagiarism:

- Enclose borrowed wording in quotation marks, and cite the publication from which the wording is taken, for example: Jones [3] states, in a similar context, that "stiction is greater than friction."
- Use a parenthetical citation for each borrowed item, even though you do not use the exact wording; for example: Jones [3] showed that under conditions of similar surface preparation but higher temperature, stiction is the dominant force, and friction is less important.

- Be sure that material you summarize or paraphrase is in your own language. Changing wording or minor sentence structure is insufficient; read what you want to paraphrase, wait a significant time, write your paraphrase without referring to the source, and check that your work does not resemble the source. Add a citation, such as "the following summary follows the reasoning of [3] ..."
- Acknowledge collaborative work. It is normal to discuss your work with others and to develop joint solutions to similar problems together. Acknowledgement of their assistance normally appears in the Acknowledgements section, unless a publication has been used or developed at the same time as your report, in which case a citation is appropriate. For example, "The data for Figure 5 was developed jointly with ..."

Conclusions: A conclusion is a conviction reached on the basis of evidence and analysis given in the body. The evidence can be of two types: the author's work as presented in the report or evidence in documents cited in the report. Every conclusion must be supported by the report itself, by cited references, or both.

Every investigation must reach a conclusion, even if it is merely that further investigation is needed or that the experiment was a failure. It should be possible to compare the Conclusions section with the Introduction section and find that every conclusion has been introduced in the Introduction section, and the outcome of every question introduced is found in the Conclusions section.

Some engineering projects result in recommendations in addition to conclusions, and it may be appropriate to use the title "Conclusions and Recommendations." The following discussion applies when a full section of recommendations is merited.

Recommendations: A recommendation answers the reader's question, "What should I do about the situation?" Give clear, specific suggestions. A recommendation, which is imperative, should not be written as a conclusion, which is declarative. The first of the following two examples is a recommendation; the second is a conclusion:

- "Improve the transfer cooling by modifying the radiation fins."
- "The transformer cooling may be improved by modifying the radiation fins."

Recommendations require decisions to be made by someone in authority; therefore, each recommendation must be specific and complete. The reader usually needs to know what is to be done, when, whether it will disrupt normal operations, how long it will take, who will do it, and the cost; or if required information is incomplete, the reader needs to know what next step should be taken.

1.3 The back matter

The back matter supports the report body, but with minor exceptions, understanding the body should not necessitate reading the back matter. The exception is this:

large or numerous figures, tables, program listings, or other material that would disrupt the flow of the body should be put into the appendices.

The Arabic page numbering of the body extends through the back matter, including the appendices, unless the appendices are complete documents with their own numbering.

References: The title of the back matter section containing the list of references is simply "References" or, sometimes, "Documents Cited."

Each document in the list of references must be uniquely identified, together with its date, authors, and sufficient information for obtaining the document. Detailed sets of rules for presenting this information have been developed for many kinds of documents. For technical documents, technical societies produce guidelines that are available in printed versions or by web search (see Section 4, Problem 1). Most of these guidelines are derived from styles developed by the American Psychological Association (APA) [5], the Modern Language Association (MLA) [6], or the University of Chicago Press [7]. These three styles are discussed in many handbooks on English writing.

The order of the entries in a reference list is normally either

- alphabetical, by last name of first author, or
- by order of citation in the text.

Each entry is given a label, which is used in the citation and is either

- numerical, in list order, or
- a combination of author name and year of publication.

Your style guide should be obeyed, but the following examples show how different document types are typically entered into a list of references:

Book:
   [1] T. K. Landauer, *The Trouble with Computers*, The MIT Press, Cambridge, MA, 1995.

Technical journal paper:
   [2] L. Peng and P.-Y. Woo, "Neural-fuzzy control system for robotic manipulators," *IEEE Control Systems Magazine*, vol. 22, no. 1, pp. 53–63, 2002.

Conference paper:
   [3] J. Fredriksson and B. Egardt, "Nonlinear control applied to gearshifting in automated manual transmissions," in *Proceedings of the 39th IEEE Conference on Decision and Control*, vol. 1, pp. 444–449. Institute of Electrical and Electronics Engineers, Sydney, 2000.

Web document:
   [4] International Bureau of Weights and Measures, *The International System of Units (SI),* Bureau International des Poids et Mesures (BIPM), Sèvres, France, 7th edition, 1998, <http://www.bipm.fr/pdf/si-brochure.pdf> (August 7, 2001).

Web documents create particular problems: the author may be difficult to identify, and the contents or web address may change over time. Therefore, show the latest date that you accessed the file, and if a printed version exists, include its particulars in preference to or in addition to the web version. There are several different guidelines for including web addresses in references; the Chicago style, recommend in [8], has been followed in this book.

The use of labels in citations is described above in Section 1.2. When the labels are numerical, the third document, for example, in a list of references, is cited as [3], (3), or by superscript^3, depending on the chosen style. Otherwise, the label consists of the author name and year of publication; [Borowick, 2000], for example.

Composing or altering a list of references by hand is problem-plagued and tedious, and these problems have influenced the guidelines mentioned above. However, the computer allows a simplified process by which document descriptions are retrieved from a central database or a local one provided by the report author, and automatically converted to the format required for the list of references. The references in this book, for example, were retrieved from a data file and formatted automatically using the "IEEE" style definition for BIBTEX, a program that produces files compatible with LATEX [9], the formatting software used for this book. Determine which system you prefer or are required to follow, and use it consistently. You may use the citations and references in this book as a model if you wish.

If you must enter all references by hand, then the easiest method is usually to list them in order of citation, with author-date labels, since adding or deleting a reference does not require changing the labels or the order of the other references.

Bibliography: A Bibliography section is included when documents that are not cited in the text are to be listed for the reader's benefit as background or further reading. The section usually begins with an introductory paragraph that describes the purpose and scope of the list. There also may be annotations (comments) included with each list entry to explain its relevance for the reader. The bibliography must be mentioned in the report text.

The bibliography format is the same as for references. The list is ordered, as appropriate, alphabetically by author, chronologically by date of publication, or by topic.

Appendices: As mentioned previously, appendices contain material that supplements the report body but is not an integral part of the text. Give each appendix a distinctive name that indicates its contents; do not simply call it Appendix A, for example, since the reader is then forced to read each appendix to see what it contains.

Large or numerous figures, tables, program listings, or other material that would disrupt the flow of the body should be put into appendices and referred to in the main text. Other typical materials are the original letter from the person requesting the report, original laboratory data, laboratory instruction sheets, engineering drawings, manufacturer's information brochures, and lengthy calculations.

2 Steps in writing a technical report

Creating a good report document requires proficiency with computer tools, a sense of design, attention to detail, and English writing skills. The possibilities for your working environment range from the desktop word processors found in small companies to elaborate project management software used in major organizations.

Desktop word processors allow you to write, revise, edit, and check what you have written. You can create or import high-quality diagrams and graphs. Spelling checkers detect misspelled words, although they cannot always detect incorrect word use. Grammar checkers also help, but they do not capture the subtleties of the English language.

In a company environment, you may have publishing software designed for producing large documents. These tools are much more robust and flexible than word processors. They can reliably assemble a document from many sources and formats and automatically do numbering, cross-referencing, tables of contents, and indexing. They provide sophisticated formatting capabilities, special characters, and powerful sub-programs to create formulas and equations. However, they may also require considerable skill to master.

Some companies not only allow the use of sophisticated writing aids, they require it. The project management software automatically saves document files as they are written and modified and controls access to them so that only one person can modify a file at any time.

Whatever your computing environment, producing a report makes repeated use of a *plan, execute, revise* cycle to create the complete document and its component parts. The following steps apply to the complete report, but with slight revision they also apply to report components:

1. Identify the reader and the purpose of the report.
2. Plan and outline the report.
3. Organize the information, and fill in the details.
4. Complete the supporting sections.
5. Revise the material until it meets the stated purpose.
6. Submit the report.

Some engineers prefer a preliminary hand-drawn list or plan for each report component; others work entirely with their computer. In either case, the computer document begins as a skeleton to which detail is added at each step. The steps are described in more detail below.

Identify the reader and the main message: Clarity and the needs of the reader are the two main considerations in report writing. Identify the reader before you begin to write. What is clear to one reader may be hopelessly opaque to another. Ideally, you will receive a memo from your boss requiring you to write a report for her or him, to answer specific questions, as part of the formal record of the progress made in an engineering project.

Without specific instructions, you must define the detailed purpose yourself. A report for your boss or a colleague in the same department requires much less background material and fewer definitions than a report for someone outside the company. In the absence of other information, write for a reader with your own general background but no specific knowledge of the special terms used in your report or the circumstances in which it is written.

Plan and outline the report: Once you know the general purpose of the report and its recipient, open the template provided by your company or a generic template that you have prepared or identified in advance. The template should contain standard report headings, but it also defines the report format: the choice of margins, point size, numbering system, front matter components, and other productions details.

Edit the title page by drafting a suitable title and entering the name of the report recipient. This page, and everything you write, is subject to later revision.

In the Introduction section, list the questions to be answered in the report, in point form and order of priority. Use this order, not just in this section, but also when presenting the details in the body of the report and when writing the Conclusions and the Recommendations sections.

Write tentative section headings in the report body, using a plan suggested for internal sections on page 88.

Organize the information, and fill in the details: When the reader and report objectives have been identified clearly, the next step is to organize the required information. If the work to be presented in the report is not complete, then your outline can assist you in setting work priorities and focussing on the questions to be answered.

Sort and organize the information as suggested on page 88. Revise the order and contents of the outline of the body as necessary.

Fill in the details of the body. Include an introductory sentence or paragraph for each section, particularly in a long report that will not be read from beginning to end at one time.

As each section is completed, write the corresponding paragraph or sub-section of the Conclusions and Recommendations sections, and collect any required appendix material.

Once the body has been completed, write the Introduction section, changing it from the list of points or rough draft that was initially written. Make sure that each conclusion in the Conclusions section corresponds to a part of the introduction.

Complete the supporting sections: Once the report body is complete, the front matter and back matter can be finished. Word processors can produce tables of contents automatically, and some are capable of formatting the documents listed in the References section automatically. Write the Summary section and the abstract, as described on page 86. Add the appendices, and number the pages consecutively with the report body.

Revise: Carefully read the report from beginning to end, making sure that the ideas are presented clearly and logically. Check the following:

1. Clarity: Are the sentences clear, concise, and correct?
2. Logic: Are the ideas developed in the right order?
3. Spelling and punctuation: Use a spell checker, but do not rely totally on it.
4. Layout and appearance: Pay special attention that diagrams are clear.

The checklist in Section 3 should be used at the final stage.

Submit the report: The final report is usually bound and accompanied by a letter of transmittal for submission to the person who requested the report. Sometimes the letter is bound into the report after the title page, but often it is simply sent with the correct number of report copies to the receiver.

The letter of transmittal is a standard business letter, such as the one shown in Figure 6.2. Near the beginning, mention the name of the report and the terms of reference or contract under which it was written. Briefly describe the purpose of the report. It may be appropriate to describe any particular points you wish to be considered as the report is read. Usually the final paragraph encourages the reader to contact the sender if further information is required.

3 A checklist for engineering reports

This is a basic checklist for use before final submittal of your report; you may wish to add to it.

1. Letter of transmittal or covering letter (page 85; Section 2):
   - ☐ State the full title of the report.
   - ☐ Refer to the original request and purpose.
   - ☐ Acknowledge all assistance, and explain the extent of the help.
   - ☐ State what parts of the report are your own work, if required.
   - ☐ Date and sign the letter.
2. Cover or binder (page 82):
   - ☐ The title and author must be visible without opening the cover.
   - ☐ The report must be securely bound.
   - ☐ There must be a neat, professional appearance.

3. Title page (page 83):
   - ☐ The title should be descriptive, specific, and sufficiently detailed for the reader to understand the report purpose.
   - ☐ Your name, organization, and address must be shown.
   - ☐ The date that the report is submitted must be shown.
   - ☐ The name of the recipient of the report must be shown.
4. Table of contents (page 86):
   - ☐ The headings and heading numbers must be the same as they appear in the report.
   - ☐ The correct page numbers must be given.
   - ☐ Important front matter headings are included: Summary, List of figures, ...
   - ☐ The full appendix titles are included (not just Appendix A, Appendix B, ...).
5. List of figures, tables (page 86):
   - ☐ List all figures and tables that appear in the body of the report.
   - ☐ Figure captions or abbreviated captions are included (not just Figure 1, Figure 2, ...).
6. Summary (page 86):
   - ☐ State the purpose or goal of the work.
   - ☐ Briefly state the methods used.
   - ☐ Briefly state the conclusions and recommendations.
7. Introduction (page 87):
   - ☐ State the purpose of the work; say what questions the report answers.
   - ☐ State the scope of the work; outline what was done to complete the work.
   - ☐ Mention the intended readership.
   - ☐ Review the history, background, or previous work.
   - ☐ Page 1 of the report is the first page of the introduction.
8. Body (page 87):
   - ☐ It must be logically organized, with numbered section and sub-section headings.
   - ☐ All figures and tables must be numbered and captioned.
   - ☐ Every figure and table must be referred to in the text.
   - ☐ Landscape figures must be readable from the right of the page.
   - ☐ Each figure or table is placed after its first reference.
9. Conclusions (page 90):

- ☐ Concisely restate the conclusions from the discussion in the detailed body sections.
- ☐ Each conclusion should correspond to a question introduced in the Introduction section.
- ☐ Do not assume that the conclusions are obvious to the reader.
- ☐ Every conclusion must be supported by the report data and analysis or by correctly cited reference documents.
- ☐ Do not introduce irrelevant comments.

10. References (page 91):
    - ☐ Every reference must be cited at least once in the body of the report.
    - ☐ Cite using a standard method, such as by numbers in brackets [3].
    - ☐ List the documents in a standard, consistent format.
    - ☐ The reference documents are listed in a standard order, such as in order of citation or alphabetically by last name of the first author.

11. Appendices (page 92):
    - ☐ Each appendix must be mentioned in the report body.
    - ☐ Appendices contain only relevant supporting material.
    - ☐ Appendices must be named descriptively (not just Appendix A, B, ...).

12. General: (Section 1.6 of Chapter 6; Chapter 7):
    - ☐ Use the active voice as much as possible.
    - ☐ Use the present or past tense but do not mix tenses.
    - ☐ Use the spell-checking feature of your word processor.
    - ☐ Proofread the report thoroughly.
    - ☐ Make sure all pages are numbered in the correct order before binding.

4 Further study

1. Use your web browser to find and bookmark style guidelines published by
   (a) The American Society of Mechanical Engineers (ASME),
   (b) The American Society of Civil Engineers (ASCE),
   (c) The American Institute of Chemical Engineers (AIChE),
   (d) The Institute of Electrical and Electronics Engineers (IEEE),
   (e) The American Institute of Physics (AIP),
   (f) The American Mathematical Society (AMS),
   (g) The Association for Computing Machinery (ACM),
   (h) The Society for Industrial and Applied Mathematics (SIAM).

2. Do a web search to locate commercial and non-commercial software tools for creating reference lists and bibliographies. Try searching for "bibliographic software."

3. Investigate the word-processing software on your computer, or the computer system provided by your institution, to find the default formats or templates for reports and other documents. How closely do they agree with the formats suggested in this text?

5 References

[1] R. Blicq and L. Moretto, *Guidelines for Report Writing*, Prentice-Hall Canada, Scarborough, ON, 4th edition, 2001.

[2] J. N. Borowick, *Technical Communication and its Applications*, Prentice Hall, Upper Saddle River, NJ, 2000.

[3] D. G. Riordan and S. E. Pauley, *Technical Report Writing Today*, Houghton Mifflin Company, New York, 1999.

[4] N. J. Higham, *Handbook of Writing for the Mathematical Sciences*, Society for Industrial and Applied Mathematics, Philadelphia, 2nd edition, 1998.

[5] American Psychological Association, *Publication Manual of the American Psychological Association*, American Psychological Association, Washington, DC, 5th edition, 2001.

[6] J. Gibaldi, *MLA Handbook for Writers of Research Papers*, Modern Language Association of America, New York, 4th edition, 1995.

[7] University of Chicago Press, *Chicago Manual of Style*, University of Chicago Press, Chicago, 14th edition, 1993.

[8] D. Hacker, *A Canadian Writer's Reference*, Nelson Thomson Learning, Scarborough, ON, 2001.

[9] L. Lamport, *LaTeX, A Document Preparation System*, Addison-Wesley Publishing Company, Reading, MA, 1994.

Chapter

Report graphics

Complex shapes and relationships are often best explained using graphics. Creating graphics is like writing: we must all understand the basic tools, but producing the best examples, such as the original of Figure 9.1, is an art.

In this chapter, you will learn:

- some principles of good graphics,
- standard formats for technical graphs,
- the simplicity and utility of straight-line graphs,
- uses for logarithmic graph scales,
- a format for including engineering calculations in documents,
- the purpose and value of sketches.

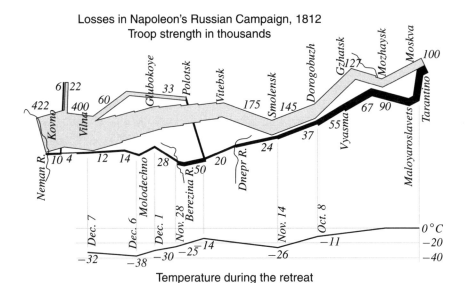

Fig. 9.1. A graphic can explain a complex subject clearly. French engineer Charles Minard drew the first version of this chart in 1861, showing French army losses caused by the march, the main battles, disastrous river crossings, and wintry temperatures. The army strength decreased from 422 000 troops to 10 000.

Chapter 9 Report graphics

1 Graphics in engineering documents

This chapter focusses on graphics in engineering reports. In this context, *graphics* are generally synonymous with *figures*; the more general graphical design elements seen in magazines and advertising are absent. Specialized graphics such as mechanical, electrical, and chemical-process design drawings are not considered here in detail; this material is often included in report appendices.

Numerous technical writing references contain advice on graphics [1, 2], but graphics is a subject by itself, discussed in [3] and [4], for example. The authors of these publications recommend that graphics be designed according to the following principles:

- clarity: the graphic must display the correct message;
- efficiency: a significant amount of data is summarized, since small amounts might be better put in a table or the text;
- balance: the graphics and text complement each other, so that each graphic is discussed in the text and reinforces it.

Often the message to be delivered by a graphic is complicated; you must have clear insight before you can decide how to represent it. Several drafts may be required to make a satisfactory diagram. Consider yourself successful if the result replaces a difficult written explanation. As an example of complex data explained with minimum detail, Figure 9.1 accomplishes the feat of relating six variables: army strength, two-dimensional position, direction of movement, date, and temperature. You might consider how long a written explanation would have to be to convey the same information, and whether it would have the same impact.

The simpler, the better, in graphics as in prose. Many computer programs are capable of producing good graphics, but beware of embellishing a diagram simply because the computer makes it easy.

Each figure in an engineering document must have a number and a caption. Tables normally have their own number sequence, separate from figures, although tables also require careful graphic design. Captions are placed below figures, but above tables. The caption should make a point; it is not simply a title. The reader should be able to understand the principal message of the figure or table from its appearance and caption without reading the report text.

2 Standard formats for graphs

Graphs are employed to show trends and functional relationships in data. The standard format for line graphs is illustrated in Figure 9.2. Basic rules for this format and other graphical figures are given in the checklist below:

1. Fit each graphic within the typeset page margins. Use a landscape page format when more detail than can be accommodated on a normal page is required. For example, Figure 9.1 might be better drawn as a landscape figure.

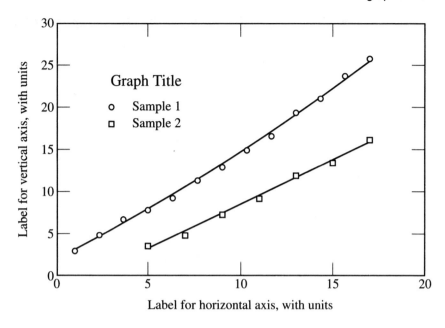

Fig. 9.2. The standard format for a line graph is shown. Each axis must have a label with units. A legend is included to identify two or more curves or for required labelling that does not appear in the caption. The graph must accurately convey the desired message as cleanly as possible.

2. Plot the independent variable of a graph along the abscissa (horizontal axis) and the dependent (observed) variables along the ordinate (vertical axis).

3. Choose the scale of each axis to include sufficient detail on the graph for the correct conclusion to be deduced. Include suitable scale marks (tic marks) for numerical scales. Figure 9.2 shows short lines at the numbers. Insert smaller scale marks between numbered scale marks when necessary. If the reader is expected to estimate the co-ordinates of points on the curves, then overlay the graph with grid lines; avoid them for clarity when the main message is the curve shape.

4. Label each axis to identify the quantity associated with it. Correct units must be included.

5. Make the graph labels readable without rotating the page. When this is impossible, make them readable when the page is rotated 90 degrees clockwise. This rule applies to left-hand pages as well as to right-hand pages when the pages are printed on both sides.

6. Mark experimental data points with distinguishing symbols when two or more sets of data are plotted. The symbols overlay any lines that are drawn to show the trend in the data. Sometimes a vertical bar is added to each

plotted point to show measurement uncertainty, as discussed in Chapters 11 and 12.

7. Include the zero value for each axis scale when the absolute value of quantities is important.

8. If the data points represent a continuous function, then draw a smooth curve through or near them. Do not draw a line if the data points do not represent a continuous function. A method for fitting a straight line to a set of points is discussed in Section 4 of Chapter 14.

2.1 Bar charts and others

Figure 9.3(a) contains a bar chart that compares the relative proportions of six quantities, arranged in order of size. Pie charts are often used for this purpose, perhaps because software for drawing them is readily available. However, comparing relative sizes of items is more difficult on pie charts than for bar charts, as shown in Figure 9.3(b); the best rule is to avoid pie charts.

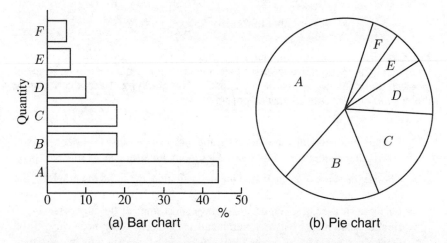

Fig. 9.3. A bar chart (a) and pie chart (b) comparing six quantities. The proportions of B compared to C and of E compared to F are difficult to distinguish on the pie chart. Bar charts are preferred.

There are many other kinds of charts, some of which are used in this book, as follows:

- Cumulative proportions, along with a bar graph, are shown in Figure 1.2.
- A flow chart, which shows a series of steps, is shown in Figure II.1.
- Organizational charts show relationships between objects, as in Figure 4.2.
- A line drawing shows essential features without extraneous detail, as in Figure 17.2.

- A geographical chart relates quantities to map position, as in Figure 9.1.
- Many other specialized drawing styles exist, such as those seen in Chapters 18 and 20.

2.2 Straight-line graphs

Many physical phenomena are modelled by functions that correspond to straight-line graphs, provided the domain is not too large. The straight line is the simplest graphical relation to understand, so it is desirable to include in a report.

If x and y are variables, and m and b are constants, then the function

$$y(x) = mx + b \tag{9.1}$$

results in a straight-line graph with $y(x)$ as abscissa and x as ordinate, as shown in Figure 9.4. The constant m is the slope of the line, and b is the vertical intercept.

Straight-line graphs such as Figure 9.4 and the lower graph in Figure 9.2 are sometimes loosely described as "linear." Avoid this term when describing a straight line; technically, the function $y(x)$ defined in Equation (9.1) is linear only if $b = 0$. However, along the bottom and left-hand scales of Figure 9.2, the distance from each scale mark to the origin is proportional to the numerical value at the scale mark, and these scales are therefore linear.

There are several ways, in addition to Equation (9.1), to describe the same straight-line function. For example, if a is the intersection of the line with the

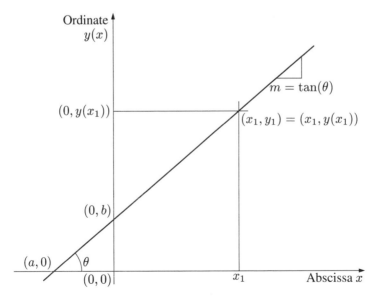

Fig. 9.4. The graph of a straight-line function is defined by two independent numbers. Two ways of defining the function are $y(x) = m\,x + b$ and $y(x) = m\,(x - a)$.

horizontal axis, then

$$y(x) = m(x - a), \qquad (9.2)$$

and if (x_1, y_1) is a point on the line, then, for all points (x, y) on the line,

$$y - y_1 = m(x - x_1). \qquad (9.3)$$

The essential fact to note is that a straight-line function is defined by two independent quantities, such as m and b. Conversely, given the straight line, only two independent numerical quantities may be derived from it.

Given n measured data pairs (x_i, y_i), $i = 1, 2, \cdots n$, it may be desired to find two quantities, such as m and b, that define a straight line passing through or near the data points in the "best" way. The lower line of Figure 9.2 is an example. There are several ways of defining the best line, but the most common method is described in Chapter 14.

2.3 Logarithmic scales

Sometimes the domain or range of a function contains both very large and very small values, and a change of variables may be desirable in order to satisfy item 3 of the checklist in Section 2. "Moore's law," shown in Figure 9.5, illustrates such a change of variables. Gordon Moore, co-founder of the Intel corporation, predicted in 1965 [6] that the number of elements in integrated circuits would

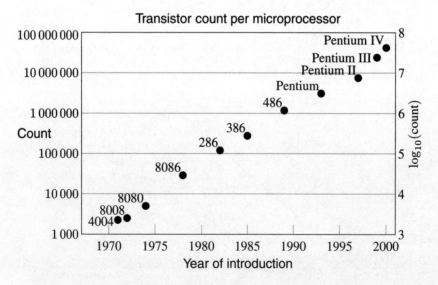

Fig. 9.5. Moore's law states that the number of elements in integrated circuits doubles every one to two years. An approximately straight line results when the vertical scale is the logarithm of the dependent variable (data from [5]).

double every one to two years. This prediction has held true for more than three decades, perhaps partly because major business decisions assumed that it would, and engineers made it happen. A linear vertical scale would make the heights of all data points in the figure, except the largest few, indistinguishably small.

The right-hand scale in Figure 9.5 illustrates the fact that if

$$y = f(x), \tag{9.4}$$

and y takes on values from 1 000 to 100 000 000, then the function

$$v = \log_{10}(y) = \log_{10}(f(x)) \tag{9.5}$$

has a compressed range from 3 to 8. The scale marks along the left-hand scale are identical to those on the right-hand scale, but the count values, rather than their logarithms, have been written, as is normally preferable. The left-hand scale is described as "logarithmic." A graph with a logarithmic vertical scale and a linear horizontal scale is said to be "log-linear." Logarithmic scales cannot show a zero value since the logarithm of zero does not exist.

A logarithmic change of variables may expand a scale rather than compress it. For example, Equation (9.5) expands the range of small values from 10^{-6} to 10^{-1} to the interval -6 to -1.

The decaying exponential function, shown in Figure 9.6, describes many physical phenomena, such as decaying radioactivity, the cooling of a hot object, decaying chemical concentration, or decaying voltage across a charged object. In these cases, the physical variable, $v(t)$, say, may be described by

$$v(t) = v_0 e^{-t/T}, \tag{9.6}$$

where v_0 is the value of v at $t = 0$, and T is called the time constant of the function. Taking the logarithm gives

$$\log_{10} v(t) = \log_{10} v_0 + \log_{10}(10^{(\log_{10} e)(-1/T)t})$$
$$= \log_{10} v_0 + (-(\log_{10} e)/T)\, t, \tag{9.7}$$

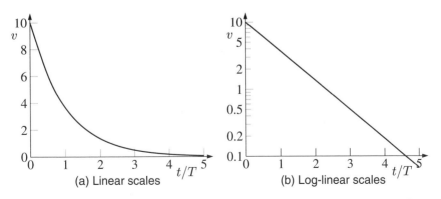

Fig. 9.6. The linear scales in (a) display the rapid approach of the function to zero better than the log-linear plot in (b).

which is a straight line on a log-linear graph, with vertical intercept v_0 and slope $-(\log_{10} e)/T$.

Figure 9.6 shows $v(t)$ plotted using linear scales and log-linear scales, for $v_0 = 10$. Notice, however, that Figure 9.6(b) may give the wrong impression; the straight line appears to the eye to decrease by half its maximum value in half of the time and to have constant slope, whereas the untransformed graph (a) on the left shows a much faster decay and varying slope. A straight-line graph, or any other, should only be used when it gives the correct visual message.

More general exponential functions can be drawn as straight-line graphs in a way that is similar to the above example. Consider the function

$$y(t-t_0) = y_0\, \alpha^{\beta(t-t_0)}, \tag{9.8}$$

where y_0, α, β, and t_0 are constants. With the substitutions $\alpha = 10^{\log_{10}\alpha}$ and $x = t-t_0$, and by taking the logarithm, we get

$$\log_{10} y(x) = \log_{10} y_0 + \beta(\log_{10}\alpha)\, x. \tag{9.9}$$

The result is a straight line on a log-linear graph, with vertical intercept y_0 and slope $\beta(\log_{10}\alpha)$.

The power function, shown in Figure 9.7, is a second common form for physical models. This function is described by the equation

$$y(x) = \alpha\, x^\beta, \tag{9.10}$$

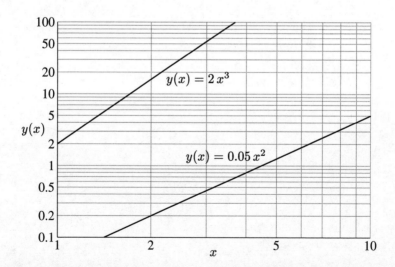

Fig. 9.7. A power function $y(x) = \alpha\, x^\beta$ becomes a straight line on a log-log plot. Two examples are shown. Logarithmic grid lines have been superimposed on the graph to allow reading the scale values of points on the line. Only scale marks should be drawn if the reader does not require such detail.

where α and β are constants. Taking the logarithm, we get

$$\log_{10} y(x) = \log_{10} \alpha + \beta \log_{10} x. \tag{9.11}$$

The function produces a straight line if $\log_{10} y(x)$ is plotted versus $\log_{10} x$ on linear scales or, alternatively, if $y(x)$ is plotted versus x on a "log-log" plot, in which both the vertical and horizontal scales are logarithmic, as in Figure 9.7.

3 Engineering calculations

Numerical and symbolic calculations are a key part of engineering practice and sometimes are crucial factors in major decisions. Therefore, calculations must be done with care, checked by others if necessary, and recorded in company documents or included in reports. Critical calculations are frequently checked by others under several circumstances, such as the following:

- For important decisions involving large expenditures of money or risk to human life, calculations are typically double-checked by a second engineer. Responsibility remains with the first engineer, but this checking provides protection against incorrect assumptions or incomplete analysis.
- Minor changes to the calculations for a project may make them applicable to a later project. The calculations and their range of validity should be easily understood by competent engineers who have not been part of the initial working group.
- In legal cases, such as inquests, civil suits, and disciplinary hearings, the engineering records of a project may be required as evidence, for the scrutiny of the court and expert witnesses. Clear assumptions and unambiguous conclusions are required to protect the calculations from challenge.

A standard format for calculations helps to ensure clarity and correctness and to guide the engineer's thought process. The format must be clear and logical, whether the calculations concern the depth of I-beams in a bridge structure, the diameter of pipes in a heat exchanger, the pitch diameter of a transmission gear, or the parameters in an electronic design. Some people have a natural talent for writing technical calculations neatly and logically; others must practise to achieve an acceptable standard.

In development laboratories, calculations are traditionally kept in bound logbooks as a permanent record; however, electronic records are now possible and increasingly used. Companies have standard formats for such files. At the end of a project, they are collected, indexed, and stored as a permanent record of the project. If project-management software is used, this process may be performed automatically.

As part of your professional education, university assignments are expected to conform to a standard format for engineering calculations. Figure 9.8 illustrates a typical problem in elementary dynamics prepared by hand in a standard format. As the figure shows, the following are standard requirements:

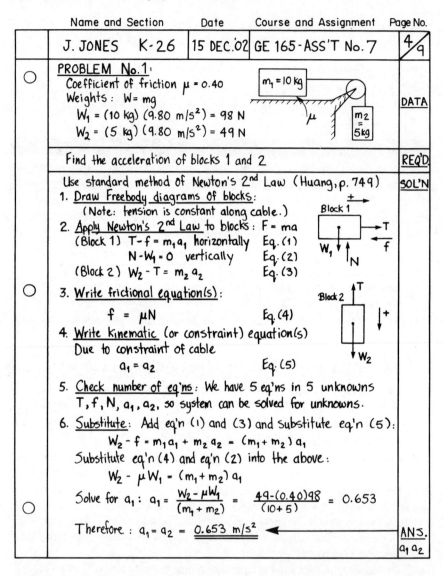

Fig. 9.8. A standard format should be used for engineering calculations. This example analyses a problem requiring elementary dynamics. Similar pages can be produced using software that will store the written material and do the calculations.

1. A space is reserved across the top for identification, including:
    - author's name,
    - date prepared,
    - project name (or course number),

- page number and number of pages in the document.
2. The left margin is sufficiently wide for binding.
3. The right margin should be wide enough to be able to flag important items.
4. A statement of the problem is required, including the given data.
5. The calculations are presented clearly. Answers must include units and should be highlighted.

Engineers use a great variety of software for tasks such as modelling, analysis, and prediction. However, as mentioned in Section 4 of Chapter 3, the professional engineer is legally responsible for the calculations and conclusions and cannot blame the computer if the output is incorrect or is misunderstood. A careful record of the data, computation, and results must be kept, containing information as follows:

- the name of the producer of the file,
- the date and time,
- the name of the software producing the result, and
- sufficient information to uniquely identify the input data.

4 Sketches

Freehand sketches, such as Figure 9.9, are made quickly, by hand, to explain ideas informally to others. A sketch can effectively communicate ideas that would require long written explanation. Engineering ideas are often sketched first, then later re-drawn as detailed engineering drawings.

In industrial design offices, the first sketch of a new idea is dated and filed; it becomes a formal record of intellectual property and may be very valuable for patent or copyright purposes.

Freehand sketches may be orthographic or pictorial, including isometric, oblique, or perspective techniques, depending on the need. The principal purpose is communication of ideas, not accuracy of lines. Of course, this is no justification for sloppiness. A talent for sketching is of great value.

Sketches are a useful record of project decisions and details and should be preserved, not destroyed. With practice, sketches can be drawn neatly and easily and slipped into a file for future reference. Many industrial research and development departments record daily intellectual effort in the form of sketches, notes, and calculations in bound lab note-books, to avoid duplication of effort if the data is lost or forgotten.

One characteristic that engineers, architects, and artists have traditionally cultivated is the ability to print alphabetic letters clearly and attractively. This ability is usually quickly acquired, since every graph, sketch, or drawing requires lettering. Texts on engineering drawing show the best method to construct each letter

110   Chapter 9   Report graphics

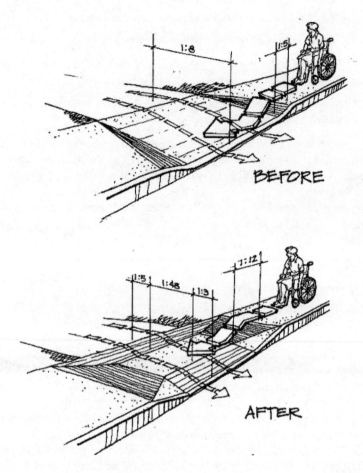

Fig. 9.9. A sketch conveys essential details informally [7] and complements or replaces long written explanation.

of the alphabet, resulting in letters that are clear and attractive with a minimum number of strokes. You may wish to progress beyond this standard lettering technique and develop an individual lettering style. This is to be encouraged, as long as the style is clear and easy to read.

5 Further study

1. Two lines have been drawn through the data points in Figure 9.2. The lower line is a straight-line graph.

(a) By reading from the lower graph, find constants m and b such that Equation (9.1) is a model for the function,

(b) Similarly, find constants m and a such that Equation (9.2) is a model for the lower graph.

(c) The upper line is not straight. Can you estimate constants b, m_1, m_2 such that the equation $y(x) = b + m_1 x + m_2 x^2$ models the data?

---

2. Suppose that the angle θ and horizontal intercept a are known in Figure 9.4. Write the formulas for the slope m and intercept b.

---

3. As described in Section 2.3, Moore's law states that the number of elements per integrated circuit doubles every T years. For the data in Figure 9.5, this law is modelled by the formula

$$n(t) = n(1971) \times 2^{(t-1971)/T}.$$

By fitting a straight line approximately to the data in the figure, determine the doubling period T that applies to the figure.

---

4. Draw the graph of the functions (a) $y(x) = 0.3\, x^{2.5}$ and (b) $y(x) = 60\, x^{-3}$ on Figure 9.7.

---

5. You are given a function $y(x) = f(x)$, where $f(x)$ is known, but is neither an exponential function such as in Equation (9.8) nor a power function such as Equation (9.10). Describe how you could choose the scales of a graph so that the function is a straight line when plotted on the graph.

---

6. Figure 9.8 illustrates a page of hand calculations in standard format. Suppose that you would like to reproduce the same page, but entirely by software, such that if you were to change any parameter, m_1, m_2, or μ, say, then the correct answer would be calculated and appear automatically on the page. See if you can find a software product or method of performing such calculations.

Answers and hints for selected problems:

1. (a) By inspection of the graph, $b = -2$, $m = 1.05$. (b) $a = -b/m$. (c) *Hint:* Draw a line tangent to the curve at the left end, which is the point x_0, y_0. Let the vertical difference between the curve and the tangent line be given by $m_2(x - x_0)^2$. The curve then has the formula $y = y_0 + m_1(x - x_0) + m_2(x - x_0)^2$, in which all constants are known. Expand the formula to obtain b.

3. Applying Equation (9.9) gives $\log_{10} n(t) = \log_{10}(n(1971)2^{-1971/T}) + ((\log_{10} 2)/T)\, t$. The coordinates for the 4004 processor are approximately (1971, 2300), and (2000, $4 \cdot 10^8$) for the Pentium IV. Thus the slope is approximately $(\log_{10} 2)/T = (\log_{10}(4 \cdot 10^8) - \log_{10}(2300))/(2000 - 1971)$. Solve for T.

5. Provided $f(\cdot)$ has an inverse function $f^{-1}(\cdot)$, then, as done for the exponential, let the vertical scale marks be drawn at positions proportional to this inverse function. Then with this choice of scales, points x, $f(x)$ on linear coordinates become $x, f^{-1}(f(x))$, producing a straight line with slope equal to 1.

6 References

[1] D. Beer, Ed., *Writing and Speaking in the Technology Professions*, IEEE Press, Piscataway, NJ, 1992.

[2] J. S. VanAlstyne, *Professional and Technical Writing Strategies*, Prentice Hall, Upper Saddle River, NJ, 1999.

[3] W. S. Cleveland, *The Elements of Graphing Data*, Wadsworth Advanced Books and Software, Monterey, CA, 1985.

[4] E. R. Tufte, *The Visual Display of Quantitative Information*, Graphics Press, Cheshire, CT, 1983.

[5] Intel Corporation, *Silicon Showcase*, Intel Corporation, 2002, <http://www.intel.com/research/silicon/mooreslaw.htm> (April 25, 2002).

[6] G. E. Moore, "Cramming more components onto integrated circuits," *Electronics*, vol. 38, no. 8, 1965.

[7] J. Barlow, P. Cannon, D. Dawson, H. D'Lil, K. Kozisek, G. Hilberry, E. Hilton, J. Markesino, C. Nagel, M. O'Connor, D. Ryan, E. Vanderslice, F. Wai, and M. Wales, *Public Rights-of-Way Access Advisory Committee Final Report*, U.S. Government, Architectural and Transportation Barriers Compliance Board, Washington, DC, 2001, <http://www.access-board.gov/prowac/commrept> (April 12, 2002).

Part III

Engineering measurements

5" Railroad Transit
Catalog No. 90

Fig. III.1. A railroad transit from the Bausch & Lomb 1908 Catalog of Engineering Instruments. Much of North America was surveyed with such instruments, and the surveys continue to be used today. (Photo courtesy of American Artifacts)

Part III. Engineering measurements

Engineering decisions, whether in design, planning, scheduling, or fabrication, are often based on measured information. However, measurements are inherently inexact. The skill and science required to make precise measurements and interpret the effect of measurement uncertainty are a basic part of your engineering education. Part III of this book introduces you to the following topics:

Chapter 10—Measurements and units: This chapter reviews the unit systems used in Canadian engineering practice. The SI (Système Internationale or International System) units are emphasized, but other systems are also described.

Chapter 11—Measurement error: Measurement, as distinct from counting, is always a form of estimation and is therefore inexact. This chapter describes the sources of measurement uncertainty and gives rules for representing inexact quantities.

Chapter 12—Error in derived quantities: How does measurement uncertainty affect a quantity derived from measurements? This chapter shows how to calculate or estimate the uncertainty in a derived quantity as a function of measurement uncertainties.

Chapter 13—Statistics: This brief introduction to the study of statistics enables you to describe and compare large sets of data using measures of central tendency (mean, mode, and median) and measures of dispersion (standard deviation and variance).

Chapter 14—Gaussian law of errors: Random errors, which are a fundamental component of measurement uncertainty, may often be described by the Gaussian probability distribution. This chapter defines the Gaussian distribution and shows how to use it in error estimation.

Chapter 10
Measurements and units

This chapter defines the commonly used engineering unit systems, explains the difference between fundamental and derived units, and describes a useful method for converting units. In it, you will learn

- the definition, components, and uncertainty of measurements;
- the elements of the unit systems used by engineers in this and other countries;
- rules for correctly writing quantities with units;
- elementary descriptions of the most commonly used unit quantities in the SI and FPS unit systems;
- the usefulness of unit algebra.

1 Measurements

Engineers frequently conduct or supervise tests and experiments that require physical measurements, such as tests of material qualities, soil and rock properties, manufacturing quality control, and experimental verification of design prototypes. A measurement that is adequate for the job at hand must be sufficiently accurate and repeatable, two properties that will be discussed in a later chapter; this chapter will concentrate on unit systems.

A *measurement* is a physical quantity that has been observed and compared to a standard quantity. The written representation of the measurement consists of two parts: a numerical value and a name, symbol, or combination of symbols that define the reference standard. For example, a distance may be measured in metres, millimetres, inches, feet, or other units, and the numerical value will depend on the unit chosen. The unit is a physical quantity that has been accepted according to experience and often international agreement as the standard by which certain measurements will be made. For example, for a time the metre was defined to be the distance between two marks on a metal bar kept in Paris.

The analysis of existing measurement techniques and the invention of new ones is an active technical art, with a community of technical specialists and regular journal publications and conferences [1], but the fundamentals are firmly fixed, except for occasional slight redefinition of the basic units.

Some units of measurement are defined in terms of others; for example, pressure may be measured in pascals. One pascal is defined as one newton per square metre, and since its definition depends on the definition of others, the pascal is called a derived unit. The units from which all others are derived are called fundamental units or base units. How many fundamental units are required? For all normally measured physical quantities, the answer is seven. Different unit systems in common use provide definitions for fundamental units and a list of other units derived from them.

The words *units* and *dimensions* are sometimes used as synonyms, but in the context of measurements, it is better to say that the dimensions of speed, say, are "distance divided by time," whereas the units of speed may be "kilometres per second" or "millimetres per year." Thus the fundamental dimensions include mass, distance, time, and other quantities corresponding to the fundamental quantities in a unit system.

2 Unit systems for engineering

The set of definitions and rules called the International System of Units, the Système International d'Unités, or simply SI [2], has been adopted in many countries by international treaty. Its base units and some derived units are listed in Table 10.1. The SI unit system is called an *absolute* system because mass is a fundamental unit and Newton's second law (force = mass × acceleration) is used to derive the force due to gravity. *Gravitational* systems adopt force as a fundamental unit and use Newton's second law to derive mass. Unit systems are classified according to whether they are gravitational or absolute and whether they use metric or English units.

In Canada and the United States, the traditional unit system is an English gravitational system, called the FPS gravitational system, in which distance, force, and time are fundamental quantities, measured in units of feet, pounds, and seconds. Both FPS and SI systems are in use in Canada, and students must be familiar with both.

The SI and the FPS systems, as well as other unit systems sometimes found in references, are compared in Table 10.2 and are discussed briefly below.

Absolute systems:

- The SI system is preferred in Canada, although quantities such as land survey measurements performed using other systems can be expected to be encountered indefinitely. Where exports, particularly to the United States, are significant, many industries continue to use versions of English or U.S. unit systems.
- The CGS system uses centimetres, grams, and seconds and was previously used extensively in science.
- The FPS absolute system is very rarely used and is not listed in the table.

Table 10.1. Base and derived SI units

Symbol	Unit Name	Quantity	Definition
m	metre, meter	length	base unit
kg	kilogram	mass	base unit
s	second	time	base unit
K	kelvin	temperature	base unit
°C	degree Celsius	temperature	(kelvin temperature) $- 273.15$
N	newton	force	$m \cdot kg \cdot s^{-2}$
J	joule	energy	$N \cdot m = m^2 \cdot kg \cdot s^{-2}$
W	watt	power	$J/s = m^2 \cdot kg \cdot s^{-3}$
Pa	pascal	pressure	$N/m^2 = m^{-1} \cdot kg \cdot s^{-2}$
Hz	hertz	frequency	s^{-1}
Electrical and Electromagnetic Units			
A	ampere	current	base unit
C	coulomb	charge	$A \cdot s$
V	volt	potential	$J/C = m^2 \cdot kg \cdot s^{-3} \cdot A^{-1}$
Ω	ohm	resistance	$V/A = m^2 \cdot kg \cdot s^{-3} \cdot A^{-2}$
S	siemens	conductance	$1/\Omega = m^{-2} \cdot kg^{-1} \cdot s^3 \cdot A^2$
F	farad	capacitance	$C/V = m^{-2} \cdot kg^{-1} \cdot s^4 \cdot A^2$
Wb	weber	magnetic flux	$V \cdot s = m^2 \cdot kg \cdot s^{-2} \cdot A^{-1}$
T	tesla	flux density	$Wb/m^2 = kg \cdot s^{-2} \cdot A^{-1}$
H	henry	inductance	$Wb/A = m^2 kg \cdot s^{-2} \cdot A^{-2}$
Dimensionless Quantities			
rad	radian	plane angle	$m/m = 1; 1/(2\pi)$ of a circle
sr	steradian	solid angle	$m^2/m^2 = 1; 1/(4\pi)$ of a sphere
mol	mole	particle count	base unit ($\simeq 6.02 \times 10^{23}$)
Light			
cd	candela	intensity	base unit
lm	lumen	flux	$cd \cdot sr$
lx	lux	illuminance	lm/m^2

Gravitational systems:

- The FPS gravitational system is the traditional system used in North America. Students must be familiar with it for a few more decades.
- The metric MKS gravitational system is rarely used. This system is not listed in Table 10.2.

Hybrid systems: The American engineering system and the European engineering system are occasionally still encountered. These are called hybrid systems, since both mass and force are fundamental units; the European hybrid units are not shown in the table.

118   Chapter 10   Measurements and units

Hybrid systems were popular for a time because units on both sides of Newton's second law equation could be arbitrarily defined. That is, the weight of an object (which is a force) could also be used as its mass. However, the use of both units, such as pound mass (lbm) and pound force (lbf) leads to confusion. To use hybrid systems, Newton's second law must be rewritten as

$$F = k_g ma, \qquad (10.1)$$

where the constant k_g equals $1/g$, where g is the acceleration of a falling object (32.2 ft/sec^2 or 9.80 m/s^2). This system should be avoided, as well as the metric hybrid unit system familiar to European engineers, mentioned above.

Table 10.2. Comparison of unit systems

	Dimensions	Absolute		Gravitational	Hybrid
		SI	CGS	FPS	American
Fundamental	Force [F]	–	–	lb	lbf
	Length [L]	m	cm	ft	ft
	Time [T]	s	sec	sec	sec
	Mass [M]	kg	g	–	lbm
Derived	Force [F]	newton kg·m/s^2	dyne g·cm/sec^2	–	–
	Mass [M]	–	–	slug lb·sec^2/ft	–
	Energy [LF]	joule N·m	erg cm·dyne	ft · lb	ft · lbf
	Power [LF/T]	watt N·m/s	erg/sec	ft · lb/sec	ft · lbf/sec
	Pressure [F/L^2]	pascal N/m^2	dyne/cm^2	lb/ft^2	lbf/ft^2

3 Writing quantities with units

Standardized rules for correctly and unambiguously representing quantities with units should be followed whenever possible. At the time of writing, there were slight differences between published standards; this book follows reference [3].

1. The symbols denoting a physical quantity consist of a number, which may be negative, followed directly by its unit symbol or symbols. The number is separated from the unit symbols by a narrow space or, if unavailable, by a normal space, except that no space is used with the symbols °, ′, ″ when they designate plane angle as degrees, minutes, and seconds, respectively:

    10.25 km   not   10.25km
    15 °C   not   15°C

2. The unit symbols must be in roman (upright) type regardless of the surrounding text. The number may be in decimal, scientific, or other notation as discussed in the next chapter. If the number is represented by a formula, then the symbols in the formula are normally in italic, or "math italic" type, with superscripts and subscripts in italic or roman type as appropriate:

   10.25 km   not   10.25 km
   abc N   not   abc N

3. SI unit symbols are lower-case, except when they are derived from the proper name of a person, in which case they normally begin with a capital letter. However, when the full name of a unit, such as a newton, is used in a sentence, it is not capitalized:

   15 kN   not   15 kn

4. The unit may be preceded by a magnitude prefix listed in Table 10.3. The prefix is regarded as an inseparable part of the symbol; thus $3 \text{ km}^2 = 3 \times (\text{km})^2 = 3 \times 10^6 \text{ m}^2$, not $3 \times 1\,000 \times (\text{m})^2 = 3\,000 \text{ m}^2$. Other computer-related magnitude prefixes that are powers of two are not separated by a space: the M in 128MB stands for 1024×1024, so $128\text{MB} = 134\,217\,728\text{B}$.

5. Unit symbols, together with prefixes if any, are treated like algebraic factors that can be multiplied or divided with numbers or other unit symbols. Multiplication between units is indicated by a half-height, or centred, dot or a half-space, division by a slash (solidus) or a negative superscript:

   $27 \text{ kg} \cdot \text{m}/\text{s}^2$ or $27 \text{ kg m}/\text{s}^2$ or $27 \text{ kg} \cdot \text{m} \cdot \text{s}^{-2}$

6. Unit symbols are not modified with subscripts:

   $V_{\max} = 200 \text{ V}$ not $V = 200 \text{ V}_{\max}$

7. It must be clear which numerical quantity and unit symbols are related:

   $3.7 \text{ km} \times 2.8 \text{ km}$ not $3.7 \times 2.8 \text{ km}$
   $87.2 \text{ g} \pm 0.4 \text{ g}$ or $(87.2 \pm 0.4) \text{ g}$ not $87.2 \pm 0.4 \text{ g}$

8. Unit symbols are not mixed with unit names in expressions, and mathematical symbols are not applied to unit names. Thus in a sentence, use kg/m^2 or kilogram per square metre, but not kilogram/m^2.

9. When long numbers are written in groups of three, the groups should be separated by a non-breaking half space rather than the comma often used in North America or the period used in some European countries:

   93 000 000 miles   not   93,000,000 miles

4 Basic and common units

Unit definitions in both SI and FPS systems are described below and compared in Table 10.2. Handbooks [4] and textbooks [5] list others. When working with

Table 10.3. Magnitude prefixes in the SI unit system by symbol, name, and numerical value.

Y	yotta	10^{24}	M	mega	10^{6}	f	femto	10^{-15}
Z	zetta	10^{21}	k	kilo	10^{3}	a	atto	10^{-18}
E	exa	10^{18}	m	milli	10^{-3}	z	zepto	10^{-21}
P	peta	10^{15}	μ	micro	10^{-6}	y	yocto	10^{-24}
T	tera	10^{12}	n	nano	10^{-9}			
G	giga	10^{9}	p	pico	10^{-12}			

different unit systems, you must be aware of the fundamental difference between absolute and gravitational systems of units and the importance of Newton's second law in defining these systems.

Force: Force is a push or a pull; it causes acceleration of objects that have mass. In the FPS gravitational system force is a fundamental unit, expressed in pounds (lb). In SI, the unit of force is the newton (N), which is the force required to give a mass of one kilogram an acceleration of one metre per second squared, derived using Newton's second law. A newton is about one-fifth of a pound or, more accurately, $1\,\text{lb} = 4.448\,222\,\text{N}$.

Length: Length or distance is a fundamental unit in all systems. The FPS unit is the foot (ft) and the SI unit is the metre (m). The United States disagrees with most other countries on the spelling, but either meter or metre is accepted when used consistently. To convert from one system to the other, by definition, $1\,\text{ft} = 0.3048\,\text{m}$.

Mass: Mass is a measure of quantity of matter. The weight of an object is the force of gravitational acceleration acting on the object; thus changes in gravitational acceleration cause a change of weight but not of mass. The kilogram is the fundamental unit of mass in SI. In the FPS gravitational system, the unit of mass is the slug, derived using Newton's second law. One slug is the mass that accelerates at one ft/sec^2 under a force of one pound: $1\,\text{slug} = 14.593\,90\,\text{kg}$.

Time: Time is a fundamental unit in all systems and is measured in seconds, abbreviated sec in the FPS gravitational system, but simply s in SI.

Pressure: Pressure has the derived units of force per unit area. In the FPS gravitational system, pressure is measured in pounds per square inch (psi) or pounds per square foot (lb/ft^2). In SI units, the unit of pressure is the pascal (Pa): $1\,\text{Pa} = 1\,\text{N/m}^2$, and $1\,\text{psi} = 6.894\,757\,\text{kPa}$.

Work and energy: Work is defined as the product of a force and the distance through which the force acts in the direction of motion. Energy has the same units and is the capacity to do work. Both are derived units in the SI and FPS systems. In the FPS gravitational system, work is measured in foot pounds (ft lb), inch pounds (in lb), horsepower hours (HP hr), British Thermal Units (BTU), and other units; the SI unit is the joule (J), and $1 \text{ ft} \cdot \text{lb} = 1.355\,818 \text{ J}$.

Power: Power is the rate of doing work. In the FPS gravitational system, power is measured in foot pounds per second (ft lb/sec) or horsepower (HP) (1 HP = 550 ft · lb/sec). In SI units, power is measured in watts (W): $1 \text{ W} = 1 \text{ J/s} = 1 \text{ N} \cdot \text{m/s}$, and $1 \text{ ft} \cdot \text{lb/sec} = 1.355\,818 \text{ W}$.

Temperature: Temperature is an indirect measure of the amount of heat energy in an object. There are four common temperature scales: Fahrenheit, Rankine, Celsius, and kelvin.

On the Fahrenheit scale used in the FPS system, water freezes at 32 °F and boils at 212 °F. Fahrenheit temperatures can be converted to Rankine by adding a constant, since the degree size is equal; t °F corresponds to $(459.67 + t)$ °R. The Rankine scale measures temperature from absolute zero and is still in common use in engineering thermodynamics in North America.

On the Celsius scale (which is a slight modification of the older centigrade scale), water freezes at 0 °C and boils at 100 °C. In fundamental SI units, water freezes at 273.15 K. The two scales have identical unit difference, so that t °C corresponds to $(273.15 + t)$ K. Absolute zero, where molecular motion stops, is the same in both Rankine and SI units (0 °R = 0 K). Note that the degree symbol (°) is not used with the kelvin symbol in SI.

Dimensionless quantities: Some measured quantities are defined as ratios of two quantities of the same kind, and thus the SI units cancel, leaving a derived unit, which is the number 1. Such measured quantities are described as being dimensionless, or of dimension 1. Many coefficients used in physical modelling and design are dimensionless; for example, the coefficient of friction is the ratio of two forces. Another example is the measurement of angle, but in this context the names "radians" and "degrees" are used for the dimensionless units. Recall that an angle in radians is the ratio of arc length over radius, and in degrees, it is the ratio of arc length divided by circumference, multiplied by the number 360, a scale factor.

U.S. and imperial units: There are differences between the U.S. and imperial units of identical name that deserve attention, particularly the units for volume. One Canadian or imperial gallon equals $4.546\,09 \times 10^{-3} \text{ m}^3$, whereas the U.S. gallon equals $3.785\,412 \times 10^{-3} \text{ m}^3$, or about 83.3 % of the imperial gallon. There are 160 imperial fluid ounces per imperial gallon but 128 U.S. fluid ounces per U.S. gallon. A summary such as [4], Appendix B of [3], or Chapter 5 of [5] should be consulted for lists of conversion factors.

5 Unit algebra

As seen in Section 3, a measured quantity is written as a number, or more generally as a formula for a number, together with the associated unit symbol or symbols. Ordinary algebra can be performed using such quantities, with the restriction that only quantities with the same units can be equated, added, or subtracted.

Consider the conversion of quantities between unit systems. Suppose, for example, that a measured power has been found to be 20 000 foot pound per minute, and this is to be converted to horsepower, which is defined as 550 ft · lb/sec. Then the written measured quantity is modified as shown:

$$20\,000 \frac{\text{ft} \cdot \text{lb}}{\text{min}} \times \frac{1\,\text{min}}{60\,\text{sec}} \times \frac{1\,\text{HP}}{550\,\text{ft} \cdot \text{lb/sec}} = \frac{20\,000}{60 \times 550}\,\text{HP} = 0.606\,\text{HP}, \tag{10.2}$$

where the original quantity has been multiplied by the number $1 = \frac{1\,\text{min}}{60\,\text{sec}}$ and again by $1 = \frac{1\,\text{HP}}{550\,\text{ft} \cdot \text{lb/sec}}$ in order to cancel the unwanted unit symbols. Sometimes, particularly when the numerical values are given by formulas rather than by pure numbers, the unit symbols are enclosed in square brackets, [,], to avoid confusion.

Unit algebra may also be performed without specifying the unit system, as in the dimensional analysis [6] of models. Let [T] represent an arbitrary unit of time, and similarly let mass be represented by [M] and length by [L]. Consider the equation from dynamics for distance s travelled by an object in time t, as the result of constant acceleration a, with initial velocity v:

$$s = vt + \frac{1}{2}at^2. \tag{10.3}$$

To derive the dimensions of s, rewrite Equation (10.3) using only the dimension symbols for each of the quantities:

$$\text{dimensions of } s = \frac{[\text{L}]}{[\text{T}]} \times [\text{T}] + \frac{[\text{L}]}{[\text{T}]^2} \times [\text{T}]^2, \tag{10.4}$$

and then perform cancellations to reduce the right-hand side to [L] + [L]. These two terms can be added since they have the same dimensions, and therefore the dimensions of s are [L], or length, as required.

The simple concept that units obey the rules of algebra as if they were numbers is a powerful check on calculations. Checking by unit conversion and cancellation that all terms of an equation have the same units, or that the dimensions are equal as in the above example, gives confidence that blunders have not been committed in developing the equation.

6 Further study

1. The equation $\sigma = My/I$ gives the stress σ at a distance y from the neutral axis of a beam subject to bending moment M. The dimensions are [F]/[L]^2 for stress, [F][L] for moment, and [L] for distance. Determine the dimensions of variable I. Can you guess what I represents, from its dimensions?

2. An object weighs 100 lbs on the earth. What would it weigh on the moon, where the acceleration of gravity is 1.62 m/s²? What is its mass on the moon?

3. The fuel consumption of vehicles is given as miles (1 mile = 1.609 344 km) per gallon in the United States, but in litres (1 L = 10^{-3} m³) consumed per 100 km in Canada. Find the formula for converting the first quantity to the second. What is 32 miles per gallon expressed in litres consumed per 100 km?

4. A direct current $i = 10$ A passes through a wire of resistance $R = 200\,\Omega$, producing heat energy at the rate $i^2 R$. How much heat is generated in three minutes? Convert the result to FPS units.

5. This is a question that uses dimensional algebra. A pendulum can be constructed from a string with length ℓ and an object with mass m. Suppose that we suspect by observation that the period T of the pendulum swing depends on length ℓ, acceleration of gravity g, and mass m. Because such relationships are generally algebraic, we conclude that

$$(\text{dimensions of } T) = (\text{dimensions of } \ell)^a \times (\text{dimensions of } g)^b$$
$$\times (\text{dimensions of } m)^c,$$

for some unknown constants a, b, and c. Determine these constants and, hence, the general relationship between the period of a pendulum and the parameters ℓ, g, and m.

Answers to problems:

1. From the given formula, $I = \frac{My}{\sigma}$; therefore, the dimensions of I are

$$\frac{[F][L] \times [L]}{[F]/[L]^2} = [L]^4,$$

that is, the dimensions of I are length to the fourth power. This parameter is a shape factor that determines how beams of different cross-section resist bending. It is called the area moment of inertia of the beam section and is calculated as $\int y^2\, dA$, where the integral is over the area of the beam cross section.

2. Using Newton's second law, $f = ma$, we have that the mass is $m = \frac{100\,\text{lb}}{32.17\,\text{ft/sec}^2} = 3.11$ slug. Mass is unaffected by gravity and is identical on the moon, where its weight will be

$$3.11\,\text{slug} \times 1.62\,\frac{\text{m}}{\text{s}^2} \times \frac{1\,\text{ft}}{0.3048\,\text{m}} = 16.5\,\text{lb}.$$

3. The units of miles per gallon are length/volume = length^{-2}, whereas the units of litres per 100 km are volume/length = length^2, so to convert a figure given in miles per gallon, we must first invert it and then use

$$1\frac{\text{gallon}}{\text{mile}} \times \frac{1\,\text{mile}}{1.609\,\text{km}} \times \frac{3.785 \times 10^{-3}\,\text{m}^3}{1\,\text{gallon}} \times \frac{1\,\text{litre}}{10^{-3}\,\text{m}^3} \times \frac{100\,\text{km}}{100\,\text{km}}$$

$$= \frac{3.785 \times 100}{1.609} \times \frac{\text{litre}}{100\,\text{km}} = 235.2\frac{\text{litre}}{100\,\text{km}}.$$

4. Energy is power × time, so the energy converted to heat is

$$(10^2 \times 200)\,\text{W} \times 3\,\text{min} \times \frac{60\,\text{s}}{1\,\text{min}} = 3.6\,\text{MJ}$$

$$= 3.6 \times 10^6\,\text{N} \cdot \text{m} \times \frac{1\,\text{lb}}{4.448\,\text{N}} \times \frac{1\,\text{ft}}{0.3048\,\text{m}} = 2.66 \times 10^6\,\text{ft} \cdot \text{lb}.$$

5. The dimensional equation is

(dimensions of T) = (dimensions of ℓ)^a × (dimensions of g)^b
 × (dimensions of m)^c,

or, substituting,

[T] = [L]^a × ([L]/[T]^2)^b × [M]^c

from which we see that $c = 0$ to eliminate the dimension of mass [M], $b = -0.5$ to make [T] a factor, and $a = -b$ to eliminate [L]. Thus the period T is proportional to $\ell^{0.5} g^{-0.5} m^0 = \sqrt{\ell/g}$. In fact, from a detailed analysis, it turns out that $T = 2\pi\sqrt{\ell/g}$.

7 References

[1] Institute of Physics, *Measurement Science and Technology*, IOP Publishing Limited, 2001, <http://www.iop.org/> (July 30, 2001).

[2] International Bureau of Weights and Measures, *The International System of Units (SI)*, Bureau International des Poids et Mesures (BIPM), Sèvres, France, 7th edition, 1998, <http://www.bipm.fr/pdf/si-brochure.pdf> (August 7, 2001).

[3] B. N. Taylor, *Guide for the Use of the International System of Units (SI)*, National Institute of Standards and Technology, Gaithersburg, MD, 1995, <http://physics.nist.gov/cuu/pdf/sp811.pdf> (February 4, 2002).

[4] T. Wildi, *Units and Conversion Charts*, IEEE Press, New York, 1991.

[5] B. S. Massey, *Measures in Science and Engineering: Their Expression, Relation and Interpretation*, Ellis Horwood Limited, Chichester, 1986.

[6] E. S. Taylor, *Dimensional Analysis for Engineers*, Clarendon Press, Oxford, 1974.

Chapter 11

Measurement error

Special words and techniques have been developed to describe and represent the inherent inexactness of physical measurement and to use inexact quantities in computations. In this chapter, you will learn

- the definition of a traceable measurement;
- a classification of uncertainties into systematic and random effects;
- correct use of the words *accuracy, precision,* and *bias;*
- how to write inexact quantities using engineering and other notation;
- the correct use of significant digits.

1 Measurements, uncertainty, and calibration

What is a measurement? One possible answer might be "A property of a physical object that can be represented using a real number." However, we have to distinguish between counting and measuring using an instrument such as in Figure 11.1. Counting is exact if the set of objects to be counted can be defined and

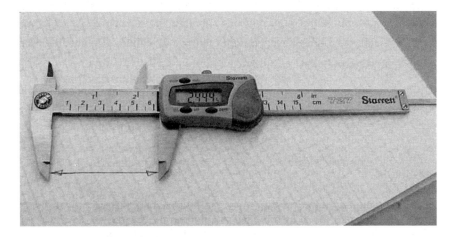

Fig. 11.1. A modern digital-readout vernier caliper, with a resolution of 0.0005 inch, or 0.01 mm.

it does not contain too many members. Thus we can count the people in an aircraft, the welds in a pipeline, and with difficulty the number of ants in a colony. However, the grains of sand on a beach are probably too many to count. Therefore let us distinguish between counting, which is exact, and measuring, which is not. The quantities that are measured are those for which base and derived units described in Chapter 10 exist.

Measured physical quantities may be inexact because the property being measured is not precisely defined. For example, to speak of the thickness of an object, such as the asphalt of a highway, is to assume two parallel plane boundaries, which do not exist when the object is looked at closely enough.

Even when it is possible to define a "true" value of a physical quantity, the instrument used to obtain its numerical value cannot be made perfectly. As illustrated in Figure 11.2, the true value is conceptually a point on the line representing the real numbers, but because of imperfect measurement, it is possible in general only to determine an estimate of the unknown true value, together with an interval in which the true value lies. The difference between the true value and the mea-

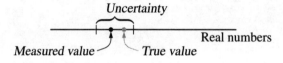

Fig. 11.2. Illustrating a true (exact) value, which is a point on the real line, and its measured estimate, together with an interval of uncertainty.

sured value is called *measurement error* or *observation error*. In this context, a measurement error does not mean a mistake or blunder. The interval determines the unknown error range, which is synonymously called the *range of uncertainty*, or simply *uncertainty*.

A measurement is fundamentally a comparison of a physical object to a standard physical object. Let us suppose that the vernier caliper of Figure 11.1 has been used to measure the thickness of an object. If the instrument has been calibrated using a set of gauge blocks manufactured for the purpose, and the gauge block manufacturer certifies their accuracy by comparison with a national standard that has been calibrated to the SI standard, then the measurement is said to be *traceable* to the international standard. In Canada, the Institute for National Measurement Standards [1] of the National Research Council provides calibration standards for factories and laboratories.

2 Systematic and random errors

Measurement errors are classified into two categories: systematic errors and random errors, although it is not always easy to identify the extent to which each is present in specific cases. The categories will be defined and illustrated.

2.1 Systematic errors

A systematic error is a consistent deviation, also called a *bias* or *offset,* from the true value. The error has the same magnitude and sign when repeated measurements are made under the same conditions. Systematic errors can sometimes be detected by careful analysis of the method of measurement, although usually they are found by calibration or by comparing measurements with results obtained independently. Three types of systematic errors are usually encountered: natural, instrument, and personal error, as follows.

Natural error arises from environmental effects. For example, temperature changes affect electronic components and measuring instruments. It may be possible to identify the effects of these phenomena, and apply a correction factor to the measurements. For example, the air buoyancy of a mass weighed by a high-precision equal-arm balance must be calculated, in order to remove its effect from the reading.

The second type of systematic error is *instrument error,* or offset, and is caused by imperfections in the adjustment or construction of the instrument. Some examples are misaligned optics, meter zero-offset, and worn bearings. In precise work, instruments are usually calibrated (checked) at several points within their range of use, and a calibration curve is included with the instrument. The calibration data permits the engineer to correct the readings for instrument errors. For example, steel surveying tapes are occasionally stretched a few millimetres during use. A stretched tape gives distance measurements that are slightly low. However, if the tape is calibrated, distance measurements can be corrected for the stretch.

The third type of systematic error is *personal error.* Such error results from habits of the observer; for example, one person may have a tendency to estimate scale values that are slightly high; another may read scale values that are low. Personal error can be reduced by proper training.

2.2 Random errors

Random errors are the result of small variations in measurements that inevitably occur even when careful readings are taken. For example, in repeated measurements, an observer may exert slightly different pressure on a micrometer, or may connect voltmeter leads in slightly different locations. Random errors from the true value do not bias the measurement, but produce both positive and negative errors with zero average value. Since the errors are random, there is no reason to favour one observation over any other, but if several observations are made, their arithmetic average is normally a better estimate of the true value than is any single measurement. The estimate improves as the number of observations increases, as discussed in Chapter 14. This technique yields an effective countermeasure against random errors: repeat and average the measurements.

3 Precision, accuracy, and bias

The terms *accuracy* and *precision* are often misused when measurements are being described. These two words are not synonyms.

The precision of a measurement refers to its repeatability. A precise measurement has small random error, and hence the discrepancies between repeated measurements taken under the same conditions are small.

A measurement that is close to the true or correct value is said to be accurate. However, "accuracy" has more than one meaning; some authors use this word to mean low total error, but others use it to mean low systematic error. When this term is used, its meaning should be made clear; otherwise use the word *bias*. To say that a measurement is unbiased is to mean unambiguously that there is no discernible systematic error. The target analogy shown in Figure 11.3 illustrates these words and their relationship to systematic and random errors. The spread

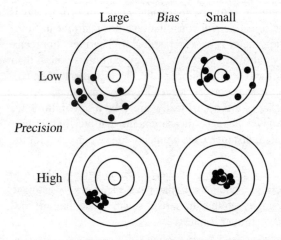

Fig. 11.3. Precision, bias, and accuracy: The two right patterns have small bias (small systematic error); the two bottom patterns have high precision (small random error). The lower right pattern is accurate; less commonly the two right patterns would be said to be accurate.

(also called scatter or dispersion) of the hits is an example of random error caused by the random motion of the shooter. The average distance between the hits and the target centre (bull's-eye) is an example of systematic error, such as might be caused by misalignment of the gun sights. Therefore, a small spread of hits located far from the centre is precise but inaccurate shooting, while a large spread around the centre is imprecise but unbiased shooting.

If unbiased measurements are repeated, their average value approaches the true value as the number of observations increases. That is, if there is no systematic error, the effect of random error can be reduced by averaging.

4 Estimating measurement error

When systematic errors are known and quantified, then the numerical observations can be corrected to remove bias. The offset is subtracted from the readings, or the instrument is recalibrated to eliminate the offset. However, the random error remains, and must be estimated and quoted with the measurement. In all cases, the technical specifications of an instrument are the first source of error estimates, but reading errors must also be considered, as follows.

Estimating the position of the needle of a gauge or meter requires interpolation between scale markings, and the precision of the result is limited by the ability of the human eye to resolve small distances. Then the precision of the readings will be a fraction of the amount between scale markings, typically one-tenth, and often this will be an adequate estimate of the measurement precision.

Readings taken from instruments with digital display must also be interpreted carefully. Typically, a four-digit numerical display will be said to have at most three and a half digits accuracy, since there can be no distinction between quantities that round to the same displayed quantity, an uncertainty of one-half of the minimum change in the displayed value. The specifications of a digital meter might indicate less precision, however. For example, a typical uncertainty for multipurpose instruments is 0.1 % of the reading plus one digit.

The precision of a measurement may become evident when the observation is repeated; any digit that changes is suspect. Include only one suspect digit in the measurement, and record it as the least significant digit in the numerical value of the measurement.

5 How to write inexact quantities

Special notation is required to show the uncertainty in the written value of an uncertain quantity. The uncertainty is normally an estimated range, and may be stated explicitly, which is always correct but sometimes inconvenient and repetitive; or implicitly, in which case care must be used both in writing and interpreting the numbers.

A quantity written using an ordinary decimal number, for example, 30 140.0, is said to be in *fixed notation*. If it is written as a decimal number with one non-zero digit to the left of the decimal, and a power of 10 is appended as a scale factor, for example $3.014\,00 \times 10^4$, it is in *scientific* notation. If the exponent of 10 is a multiple of 3 to correspond to the SI prefixes, for example, $30.140\,0 \times 10^3$, the number is in *engineering* notation.

5.1 Explicit uncertainty notation

Sets of measured values are often listed together in tables. When the uncertainties are identical, the uncertainty may be specified in a caption or table heading. Otherwise, the uncertainty must be given with each measurement. The measurement may be expressed in fixed, scientific or engineering notation, and the uncertainty may be given in either absolute or relative form, as will be described.

In fixed notation, the numerical value of the measurement is written together with an estimate of the half-range, when the range is symmetrical about the measured value as is often true. For example, a certain measurement of the speed of sound v_s in air at 293 K gives

$$v_s = 343.5\,\text{m/s} \pm 0.9\,\text{m/s} = (343.5 \pm 0.9)\,\text{m/s} \tag{11.1}$$

where the estimated maximum error is 0.9 m/s. The unit symbol m/s is treated as an algebraic factor multiplying the contents of the parentheses. Rather than indicating the *absolute uncertainty* as above, the *relative uncertainty* may be written as a percentage of the magnitude. In this case, $(0.9\,\text{m/s})/(343.5\,\text{m/s}) = 0.26 \times 10^{-2} = 0.26\,\%$, so that the measured speed can be written as

$$v_s = 343.5\,(1 \pm 0.26\,\%)\,\text{m/s}. \tag{11.2}$$

The unit symbol % simply means the number 0.01 when used strictly (see Chapter 11 of [2]), but the ambiguous notation $343.5\,\text{m/s} \pm 0.26\,\%$ is often accepted. Also, some published standards omit the space before the percent sign.

Using scientific notation, the above measurement is

$$v_s = (3.435 \pm 0.009) \times 10^2\,\text{m/s} \tag{11.3}$$

which means that v_s lies between 3.426×10^2 m/s and 3.444×10^2 m/s. The uncertainty is written with the same scale multiplier as the value of the measurement. In engineering notation, the above speed is

$$v_s = (343.5 \pm 0.9) \times 10^0\,\text{m/s}. \tag{11.4}$$

5.2 Implicit uncertainty notation

If a quantity is written without an explicit uncertainty, the uncertainty is taken to be ±5 in the digit immediately to the right of the least significant digit. For example, the statement $t = 302.8$ K is equivalent to $t = (302.8 \pm 0.05)$ K. However, in fixed notation, ambiguity may result, as shown in the next section.

6 Significant digits

The significant digits of a written quantity are those that determine the precision of the number. It is generally a grave mistake to record a quantity to more significant digits than justified by the measurement, with the exception that one or more suspect digits is sometimes retained when the effect of random errors will be reduced by averaging as discussed in Section 6.2.

Relative precision is unaffected by the power-of-10 scale factors in scientific or engineering notation. Significant digits are identified as follows:

- All non-zeros are significant, and all zeros between significant digits are significant.
- Leading zeros are not significant.
- Trailing zeros are significant in the fractional part of a number.

In fixed notation, trailing (right-hand) zeros in a number without a fractional part may cause ambiguity. Therefore, if right-hand zeroes should be considered to be significant, explicit mention of the fact must be made, or the uncertainty must be included. The following are some examples of numbers in fixed notation:

0.0350 oz	3 significant digits
90 000 000 miles	1 significant digit, ambiguous
92 900 000 miles	3 significant digits, ambiguous
(92 900 000 ± 500) miles	5 significant digits

In scientific notation, the integer part is normally one non-zero digit, and the number of significant digits is one more than the number of digits after the decimal point. To illustrate, the distance from the earth to the sun is written as shown:

9×10^7 miles	1 significant digit
9.3×10^7 miles	2 significant digits
9.290×10^7 miles	4 significant digits

In engineering notation, all of the digits except leading zeros are considered to be significant, and the exponent of 10 in the scale factor is a multiple of 3. In engineering notation, the above distances would be written:

0.09×10^9 miles	1 significant digit
93×10^6 miles	2 significant digits
92.90×10^6 miles	4 significant digits

6.1 Rounding numbers

When a number contains digits that are not significant, rounding is used to reduce the number to the appropriate number of significant digits. Rounding should not be confused with truncation, which means simply dropping the digits to the right of a certain point. Rounding drops digits, but may increase the rightmost retained digit by 1, according to the following rules:

When the digits to be discarded begin with a digit less than 5, then the retained digits are unchanged; for example, rounding to three or two digits,

$$3.234 \rightarrow 3.23 \text{ or } 3.2$$
$$9.842 \rightarrow 9.84 \text{ or } 9.8. \tag{11.5}$$

When the digits to be discarded begin with a digit greater than 5, or with a 5 that has at least one non-zero digit to its right, then the rightmost retained digit is increased by 1; for example, rounding to three or two digits,

$$3.256 \rightarrow 3.26 \text{ or } 3.3$$
$$9.747 \rightarrow 9.75 \text{ or } 9.7 \text{ but } not\ 9.8. \tag{11.6}$$

When the digits to be discarded begin with a 5 and all following digits are zero, then the rightmost retained digit is increased by one if it is odd; otherwise it is left unchanged. This means that the resulting final digit is always even; for example, rounding to two digits,

$$\begin{aligned} 3.25 &\to 3.2 \\ 3.250 &\to 3.2 \\ 3.350 &\to 3.4. \end{aligned} \tag{11.7}$$

6.2 The effect of algebraic operations

When computations are performed using inexact measured quantities, the results are also inexact. The quantities are entered to the number of digits justified by experimental error, and computations are performed to the full precision of the computer. However, the final computed value must not be written using more significant digits than are justified; it must be rounded to imply the correct interval of uncertainty. Basic rules will be given for writing the result of simple operations, anticipating the more detailed analyses in Chapters 12 to 14.

Addition: Let x and y be two measured values, with (unknown) measurement errors Δx, Δy, respectively, in the ranges given by implicit notation. Then the sum of the true values is

$$(x + \Delta x) + (y + \Delta y) = (x + y) + (\Delta x + \Delta y) \tag{11.8}$$

where $x+y$ is the calculated sum and $\Delta x + \Delta y$ the resulting error. The magnitude of the error in the sum is $|\Delta x + \Delta y|$, corresponding to an absolute uncertainty which is at most twice that of the least precise operand. Therefore, for a small number of additions or subtractions, the result is often rounded to the absolute uncertainty of the least precise operand.

Subtraction: The analysis for addition shows that the absolute uncertainty can at most double with each addition. This conclusion is similarly true for subtraction, but the *relative* uncertainty, and hence the number of significant digits, may change drastically. For example, $5.75 - 5.73 = 0.02$ with an implied error of ± 0.01, so the result can only be written to one significant digit at most, rather than the three digits of the operands. The following examples illustrate addition and subtraction:

4.16	0.123	6.162	25.4
-12.3214	-178	-12.3214	3.1416
91.2	0.002164	6.150	0.3183
83.0	-178	-0.009	28.9

Multiplication and division: The relative uncertainty in a product or ratio is typically at most the sum of the relative uncertainty of the factors. This result is shown for the product as follows; a similar analysis holds for division. Given two quantities $x + \Delta x = x(1 + \Delta x/x)$ and $y + \Delta y = y(1 + \Delta y/y)$ as before, the relative errors are $\Delta x/x$ and $\Delta y/y$, often expressed in percent. Then the computed product of the two measured values x and y is simply xy, but including the errors in the computation gives

$$(x + \Delta x)(y + \Delta y) = xy + y\,\Delta x + x\,\Delta y + \Delta x \Delta y$$
$$= xy\left(1 + \frac{\Delta x}{x} + \frac{\Delta y}{y} + \frac{\Delta x}{x}\frac{\Delta y}{y}\right). \tag{11.9}$$

Typically the relative errors $\Delta x/x$ and $\Delta y/y$ are small, so the rightmost term in the parentheses can be ignored, and the relative error of the product is at most approximately twice the largest relative uncertainty of the factors. Consequently, the result of a multiplication is often rounded to the number of significant digits of the factor with the fewest significant digits. The following examples illustrate this rule:

$$\begin{aligned} 2.6857 \times 3.1 &= 8.3 \\ (489.5)^2 &= 239\,600 \\ 236.52/1.57 &= 151 \\ 25.4 \times 0.866\,025 &= 22.0 \end{aligned} \tag{11.10}$$

Caution! The rules for single operations are not applicable in general to more complex calculations. Consider the product of four measured values, each given to two significant digits, implying an uncertainty of ± 0.05:

$$p = 1.1 \times 1.2 \times 1.3 \times 2.4 = 4.1184. \tag{11.11}$$

Rounded to two digits according to the precision of the individual factors, the result would be written as $p = 4.1$, which implies that $4.05 \leq p \leq 4.15$. However, in the worst case, when the errors have the same sign and maximum magnitude, the actual range minimum and maximum are given by the calculations

$$\begin{aligned} 1.05 \times 1.15 \times 1.25 \times 2.35 &= 3.5470, \\ 1.15 \times 1.25 \times 1.35 \times 2.45 &= 4.7545; \end{aligned} \tag{11.12}$$

that is, $3.5470 \leq p \leq 4.7545$. The result should be written as $p = 4$, implying the range $3.5 \leq p \leq 4.5$, which approximates the above range. However, judgement may be required about whether a worst-case analysis is appropriate.

Arithmetic mean: The arithmetic mean, or average, of a set of repeated measurements of a single quantity subject to random errors is more precise than an individual measurement. Anticipating the discussion in Chapter 14, the uncertainty in the mean of N observations is $1/\sqrt{N}$ of the uncertainty in one observation. Each additional significant digit appended to the mean value requires the uncertainty to be reduced by a factor of 10, so adding m digits implies that

$$\frac{1}{\sqrt{N}} = \left(\frac{1}{10}\right)^m \tag{11.13}$$

134 Chapter 11 Measurement error

from which $m = 0.5 \log_{10} N$. Converting m to integer values gives the table:

N	1 to 9	10 to 99	10^2 to 999	10^3 to 9999	10^4 to 99999	$\cdots$
m	0	1	1	2	2	$\cdots$

Thus no digits should be added for an average of less than 10 measurements, one for 10 to 999, and so on, or perhaps more conservatively, depending on the uncertainty of individual measurements.

7 Further study

1. A dinosaur skeleton was discovered in 1990 and estimated to be 90 000 000 years old. Does this mean that the skeleton was 90 000 010 years old in the year 2000? (Adapted from [3])

2. The original observations by Dr. Wunderlich of the temperature of the human body were averaged and rounded off to 37 °C, which was the nearest Celsius degree. Therefore, the precision of the measurement was presumably ±0.5 °C. When this measurement was converted to the Fahrenheit scale, it became 98.6 °F. What is the correct interpretation of the tolerance on the Fahrenheit figure?

Recently, extensive measurements of body temperature gave a mean of only 98.2 °F. Convert this to an equivalent kelvin temperature with the proper tolerance. (Adapted from [3])

3. The notation 343.5 m/s ± 0.26 % was described as ambigous on page 130. What rule about units does it break?

8 References

[1] A. Alberto, *Institute for National Measurement Standards*, National Research Council Canada, 2001, <http://www.nrc.ca/inms/inmse.html> (August 10, 2001).

[2] B. N. Taylor, *Guide for the Use of the International System of Units (SI)*, National Institute of Standards and Technology, Gaithersburg, MD, 1995, <http://physics.nist.gov/cuu/pdf/sp811.pdf> (February 4, 2002).

[3] J. A. Paulos, *A Mathematician Reads the Newspaper*, Harper Collins, New York, 1995.

Chapter 12

Error in derived quantities

As we saw in the last chapter, calculations performed with inexact data give inexact results. Some measured variables may contribute more to the uncertainty in the final result than others, and it might be useful to improve the measurement procedures for those variables. Hence, the key questions are: what is the uncertainty in the calculated result and which measured data items are most significant in contributing to the uncertainty in the result? Furthermore, the uncertainties of the arguments may not be given as absolute values or ranges, but as relative quantities. The requirement may then be to find the relative uncertainty in the calculated result, and the measured items that contribute most to it.

Using elementary mathematical notation, let us suppose that the function $f(x, y, z, \cdots)$ of the variables $x, y, z, \cdots$, is to be calculated. Let the measured values of the arguments be $x_0, y_0, z_0, \cdots$ respectively, so that the nominal computed value is $f(x_0, y_0, z_0, \cdots)$. Then we must calculate the range of f given the ranges of its arguments and the sensitivity of f to variations of its arguments from their measured values. In this chapter, you will learn

- the direct method of calculating the exact range of a function when the measurement error range is known,
- how to approximate the worst-case range for more complex calculations,
- how to approximate the range when estimates of typical measurement errors are known.

1 Method 1: Exact range of a calculated result

The first method calculates uncertainty by calculating the range of the final result, using the range of the input variables.

Assumptions:

1. The exact range of the computed quantity $f(\cdots)$ is required.
2. The exact ranges of the arguments $x, y, z, \cdots$ are known.

The above assumptions are not always strictly true, since the range of a measured quantity is usually an estimate rather than an exact quantity. However, this basic method requires no approximations as do the other methods.

Chapter 12 Error in derived quantities

Example 1. Measurement of resistance.

The resistance in ohms (Ω) of an element in an electric circuit is to be calculated, given the measured values $V = 117\,\text{V}$ of the voltage across the element and $I = 2.23\,\text{A}$ of the current through the element. The resistance of the element is defined to be $R(V, I) = \frac{V}{I}$. Assume that the ranges of I and V are

$$I_{\min} \leq I \leq I_{\max} \quad \text{and} \quad V_{\min} \leq V \leq V_{\max}$$

where $I_{\min}$, $I_{\max}$, $V_{\min}$, and $V_{\max}$ are known positive values. Then the nominal resistance is the value

$$R = \frac{V}{I} = \frac{117}{2.23} = 52.5$$

and for this simple function the range of the calculated resistance is seen to be

$$\frac{V_{\min}}{I_{\max}} \leq R \leq \frac{V_{\max}}{I_{\min}}.$$

Using the extreme values for V and I implied by their significant digits,

$$\frac{V_{\min}}{I_{\max}} = \frac{116.5}{2.235} = 52.13 \leq R \leq 52.81 = \frac{117.5}{2.225} = \frac{V_{\max}}{I_{\min}}.$$

Notice that the nominal value of R is not at the mid-point of its calculated range.

Example 2. Mass of a conical volume of crushed glass.

A quantity of fine crushed glass for recycling has been piled in an approximate cone of diameter d and height h, and its mass is to be calculated from its density ρ. The volume of a cone is one-third the height times the area of the base, so the mass is given by the formula $m(\rho, h, d) = \frac{\pi}{12}\rho h d^2$. The measured values are $d = 1.73\,\text{m} \pm 0.05\,\text{m}$ and $h = 1.03\,\text{m} \pm 0.05\,\text{m}$ where the uncertainty arises since the pile is not perfectly circular or conical. The density of fine crushed glass is given as $1120\,\text{kg/m}^3 \pm 30\,\text{kg/m}^3$.

The nominal numerical value of the mass is

$$m = \frac{\pi}{12} \times 1120 \times 1.03 \times 1.73^2 = 904.$$

The ranges of the arguments of the function $m(\rho, h, d)$ are

$$1090 \leq \rho \leq 1150, \quad 0.98 \leq h \leq 1.08, \quad 1.68 \leq d \leq 1.78,$$

and since m is proportional to ρ, h, and d^2, its minimum and maximum occur at the minimum and maximum respectively of its arguments, so that

$$m \geq \frac{\pi}{12} \times 1090 \times 0.98 \times 1.68^2 = 789, \quad \text{and}$$

$$m \leq \frac{\pi}{12} \times 1150 \times 1.08 \times 1.78^2 = 1030.$$

Again the nominal calculated value does not occur at the mid-point of the range.

It should not be concluded from the above two examples that the extreme values of the range of a function always occur at the extreme values of the ranges of its arguments. Figure 12.1 illustrates a function $f_1(x)$ of one argument x for which the range is not obtained by calculating the function at maximum or minimum values of x. However if, over the range of x, the function is bounded and monotonic, that is, always non-decreasing or always non-increasing, as illustrated by $f_2(x)$ in the figure, then its extreme values are reached at the extreme values of x. If a function of several variables has this property with respect to each of its arguments, as is true for differentiable functions and sufficiently small argument ranges, then the range of the function is relatively simply found by searching at the extreme values of the arguments.

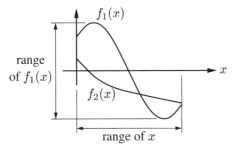

Fig. 12.1. Computing $f_1(x)$ at the upper and lower boundaries of x does not give the range of $f_1(x)$ over the range of x. The extremes of $f_2(x)$, which is monotonic and more typical for small ranges of x, are reached at the extremes of x.

2 Method 2: Linear estimate of the error range

The second method for calculating the range of a function of inexact variables is an approximation to the first method. An approximation is often sufficient since the range of measured quantities is typically estimated rather than known exactly, and further approximation does not change the character of conclusions based on the measurements. It turns out that the magnitude of the uncertainty in the calculated value is the sum of the effects of the errors in the variables. This method provides an essential result that will also be used in the third method.

Assumptions:

1. The range calculated in the previous method is to be *approximated*.
2. The function $f(x, y, z, \cdots)$ is differentiable at the measured values x_0, y_0, z_0, $\cdots$ of its arguments.

3. The measurement errors in the measured values are independent of each other, and may be either positive or negative.
4. The ranges of the measured values are known.

For simplicity, consider first a function $f(x)$ of one variable, as illustrated in Figure 12.2. At nominal (i.e., measured) value $x = x_0$, $f(x)$ has the value $f(x_0)$ and slope $\left.\dfrac{df}{dx}\right|_{x=x_0}$, assuming that the derivative exits at x_0. Then from the definition of the derivative, for small changes Δx in the measured value, the function changes by an amount Δf given approximately by

$$\Delta f \simeq \left.\frac{df}{dx}\right|_{x=x_0} \Delta x. \tag{12.1}$$

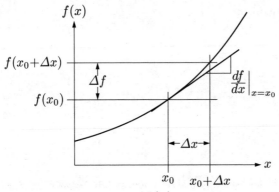

Fig. 12.2. The change in $f(x)$ is approximately the slope of f at x_0 multiplied by the deviation Δx.

Example 3. Uncertainty calculation.

Suppose that the quantity $f(x) = 1 + x^2$ is to be calculated, and that the measured value of x is $x_0 = 0.64$, implying an uncertainty of ± 0.005 and a nominal value $f(x_0) = 1 + 0.64^2 = 1.41$. Differentiating the formula for $f(x)$ gives the formula

$$\frac{df}{dx} = 2x,$$

but when this formula is evaluated at the measured value, we get

$$\left.\frac{df}{dx}\right|_{x=x_0} = 2x|_{x=0.64} = 1.28$$

which is not a formula, but a constant. Then the uncertainty in the calculated quantity is approximately

$$\Delta f \simeq 1.28 \times 0.005 = 0.0064.$$

2.1 Sensitivities

Expressing the result of the last section in words, the approximation illustrated in Figure 12.2 for the change in $f(x)$ near the point of tangency is

$$\text{change in } f(x) \simeq (\text{rate of change with respect to } x) \times \Delta x. \tag{12.2}$$

To extend the above result to a function of two variables, imagine a three-dimensional picture showing the surface $f(x, y)$ as a function of x and y, with a plane tangent to the surface at the point x_0, y_0, $f(x_0, y_0)$. In moving from the tangent point to a nearby point on the surface, the f-distance moved is approximately the slope of the plane measured in the x-direction times the x-distance moved, plus the slope in the y-direction times the y-movement. A similar argument holds for functions of several variables, as follows:

change in $f(x, y, z, \cdots) \simeq$

(rate of change with respect to x) $\times \Delta x$

$+$ (rate of change with respect to y) $\times \Delta y$

$+$ (rate of change with respect to z) $\times \Delta z$

$+ \cdots$ $\tag{12.3}$

for small Δx, Δy, Δz, $\cdots$, provided that all the required quantities exist and are unique. Rewriting this approximation more symbolically, we have

$$\Delta f \simeq S_x \Delta x + S_y \Delta y + S_z \Delta z + \cdots \tag{12.4}$$

where the rates of change of f with respect to its arguments are *sensitivity constants*, or simply *sensitivities*, calculated as follows:

The sensitivity constant S_x is obtained by calculating the formula for the derivative of f with respect to x, keeping y, z, $\cdots$ constant, and evaluating the formula at the measured values $x = x_0$, $y = y_0$, $z = z_0 \cdots$

The sensitivity constant S_y is obtained by calculating the formula for the derivative of f with respect to y, keeping x, z, $\cdots$ constant, and evaluating the formula at the measured values $x = x_0$, $y = y_0$, $z = z_0 \cdots$

Further constants are calculated similarly. The derivatives calculated above are called *partial derivatives* of f, and are written $\frac{\partial f}{\partial x}$, $\frac{\partial f}{\partial y}$, $\cdots$ instead of $\frac{df}{dx}$, $\frac{df}{dy}$, $\cdots$, to indicate that differentiation is performed with respect to the indicated variable while keeping the others constant.

Example 4. Sensitivity constants for Example 1, resistance.

In Example 1, the formula for the computed result was

$$R(V, I) = \frac{V}{I},$$

so that the required derivatives and associated sensitivities are

$$\frac{\partial R}{\partial V} = \frac{1}{I}, \quad S_V = \frac{1}{I}\bigg|_{\substack{V=117 \\ I=2.23}} = \frac{1}{2.23} = 0.448,$$

$$\frac{\partial R}{\partial I} = \frac{-V}{I^2}, \quad S_I = \frac{-V}{I^2}\bigg|_{\substack{V=117 \\ I=2.23}} = \frac{-117}{2.23^2} = -23.5.$$

Thus the computed resistance will change approximately by 0.448 per unit voltage change, and by -23.5 per unit current change.

Example 5. Sensitivity constants for Example 2, mass of conical volume.

In Example 2, the formula for the computed result was

$$m(\rho, h, d) = \frac{\pi}{12}\rho h d^2,$$

so that the sensitivities are calculated as

$$\frac{\partial m}{\partial \rho} = \frac{\pi}{12}hd^2, \quad S_\rho = \frac{\pi}{12}hd^2\bigg|_{\substack{\rho=1120 \\ h=1.03 \\ d=1.73}} = 0.807,$$

$$\frac{\partial m}{\partial h} = \frac{\pi}{12}\rho d^2, \quad S_h = \frac{\pi}{12}\rho d^2\bigg|_{\substack{\rho=1120 \\ h=1.03 \\ d=1.73}} = 878,$$

$$\frac{\partial m}{\partial d} = \frac{\pi}{6}\rho h d, \quad S_d = \frac{\pi}{6}\rho h d\bigg|_{\substack{\rho=1120 \\ h=1.03 \\ d=1.73}} = 1045.$$

2.2 Relative sensitivities

The sensitivity constants defined in Section 2.1 allow an answer to one of the questions posed at the beginning of this chapter: for small measurement errors, the measurements that potentially contribute most to changes in the calculated result are those that have the largest sensitivity constants. However, a modified question may be posed: what relative change in f results from small relative changes in its arguments, and for relative changes, which variables may have the largest effect on the computed result?

For a simplified notation, write f_0 in place of $f(x_0, y_0, z_0, \cdots)$, and divide Equation (12.4) by f_0, so that the left side $\frac{\Delta f}{f_0}$ is the relative change in the computed result f_0, assuming that f_0 is non-zero. The result can be written

$$\frac{\Delta f}{f_0} \simeq \frac{S_x \times x_0}{f_0}\left(\frac{\Delta x}{x_0}\right) + \frac{S_y \times y_0}{f_0}\left(\frac{\Delta y}{y_0}\right) + \frac{S_z \times z_0}{f_0}\left(\frac{\Delta z}{z_0}\right) + \cdots$$

(12.5)

where $\frac{\Delta x}{x_0}$ is the relative change in x, and so on for the other variables. The constants on the right-hand side of the form $\frac{S_x \times x_0}{f_0}$, and similarly for the other variables, are *relative sensitivity coefficients*. The largest such coefficient corresponds to the argument for which a relative change produces the largest relative deviation of f from f_0.

Example 6. Relative sensitivities for Example 4, resistance.

Example 4 investigated the influence of changes in measured values on changes in calculated resistance. In terms of relative changes, the result looks somewhat different. For the formula for $R(V, I)$, the relative sensitivity constants are

$$\frac{S_V \times V_0}{R(V_0, I_0)} = \frac{0.448 \times 117}{52.5} = 1.00,$$

$$\frac{S_I \times I_0}{R(V_0, I_0)} = \frac{-23.5 \times 2.23}{52.5} = -1.00.$$

Thus, a change of one percent in either V or I will result in approximately one percent change in the calculated resistance, and the two measured values V and I are equally important in terms of relative changes. The computed coefficients have a suspicious simplicity, which will become clear in Section 2.4.

2.3 Approximate error range

The approximation for the change in the computed function $f(x, y, z, \cdots)$ was given in Equation (12.4), where it was assumed that the deviations Δx, Δy, Δz, $\cdots$ are independent and may be of either sign. The calculated constants S_x, S_y, S_z, $\cdots$ also may be of either sign. To approximate the worst-case deviation $|\Delta f|_{\max}$, take absolute values to get the formula

$$|\Delta f|_{\max} \simeq |S_x||\Delta x| + |S_y||\Delta y| + |S_z||\Delta z| + \cdots, \tag{12.6}$$

which is an approximation to the exact result discussed in Section 1. This formula is particularly simple to use if the sensitivity constants can be obtained easily.

Example 7. Computed range for Example 4, resistance.

From the values give in Examples 1 and 4, the approximate range of the computed result is

$$|\Delta R|_{\max} \simeq |S_V||\Delta V| + |S_I||\Delta I|$$
$$= |0.448| \times |0.5| + |-23.5| \times |0.005| = 0.3,$$

giving the calculated resistance as

$$R = 52.5 \pm 0.3,$$

which may be compared with the exact result given in Example 1.

Example 8. Computed range for Example 5, mass of conical volume.

The numerical values from Examples 2 and 5 give

$$|\Delta m|_{\max} \simeq |S_\rho||\Delta \rho| + |S_h||\Delta h| + |S_d||\Delta d|$$
$$= |0.807| \times |30| + |878| \times |0.05| + |1045| \times |0.05|$$
$$= 120,$$

giving the calculated mass as $m = 904 \pm 120$, which may be compared with the exact result of Example 2.

2.4 Application of Method 2 to algebraic functions

The quantities computed from measured data are often simple algebraic functions of their arguments. Examples of such simple functions will be investigated here.

Linear combinations of measured values: Simple addition and subtraction are special cases of functions that are linear combinations of variables with constant coefficients, for example,

$$f(x, y, z) = ax + by - cz, \tag{12.7}$$

with constant a, b, c.

For this simple function, the exact range is given by

$$|\Delta f|_{\max} = |a||\Delta x| + |b||\Delta y| + |c||\Delta z|, \tag{12.8}$$

that is, absolute values of each term are added.

Method 2 requires sensitivity constants, but in this case these are simply the constant coefficients of the variables in the expression, so that

$$\Delta f = S_x \Delta x + S_y \Delta y + S_z \Delta z = a\Delta x + b\Delta y - c\Delta z, \tag{12.9}$$

giving exactly the same result as Method 1:

$$|\Delta f|_{\max} = |a||\Delta x| + |b||\Delta y| + |c||\Delta z|. \tag{12.10}$$

The function is said to be linear in each of its arguments, and no approximation is required.

In the special cases of addition and subtraction, simply add the absolute error magnitudes as in Section 6.2 of Chapter 11; that is, if

$$f(x, y) = x + y \tag{12.11}$$

then

$$|\Delta f|_{\max} = |1||\Delta x| + |1||\Delta y| = |\Delta x| + |\Delta y|. \tag{12.12}$$

Example 9. Addition and subtraction.

$$\left\{\begin{array}{l} x = 1.5 \pm 0.1 \\ y = 2.3 \pm 0.2 \end{array}\right\} \to x + y = 3.8 \pm 0.3 \text{ and } x - y = -0.8 \pm 0.3$$

Multiplication (or division): For these two operations, the rule is to add relative uncertainties, as in Section 6.2 of Chapter 11, rather than absolute uncertainties. If a product is written as

$$f(x, y) = xy, \tag{12.13}$$

then its sensitivity with respect to x is simply the value of y, and vice versa, so that

$$|\Delta f|_{\max} \simeq |y||\Delta x| + |x||\Delta y|. \tag{12.14}$$

Dividing both sides by $|f(x, y)| = |x||y|$ to obtain the relative uncertainty of the result gives

$$\frac{|\Delta f|_{\max}}{|f|} \simeq \frac{|y||\Delta x|}{|x||y|} + \frac{|x||\Delta y|}{|x||y|} = \frac{|\Delta x|}{|x|} + \frac{|\Delta y|}{|y|}. \tag{12.15}$$

A similar derivation produces the above result for simple division.

Example 10. Multiplication.

$$\left\{\begin{array}{l} x = 1\,(1 \pm 1\,\%) \\ y = 2.3\,(1 \pm 2\,\%) \end{array}\right\} \to x \cdot y = 2.3\,(1 \pm 3\,\%) \text{ and } x/y = 0.43\,(1 \pm 3\,\%).$$

Algebraic terms: Consider the term

$$f(x, y, z) = A \frac{x^a y^b}{z^c} = A x^a y^b z^{-c} \tag{12.16}$$

where A, a, b, and c are constant.

In contrast to the example of a linear combination of variables, general conclusions cannot be reached about the application of Method 1 to this function, since the range depends on the numerical values of the quantities in f; for example, if $z = 0.05 \pm 0.1$, then for some values of z in this range, f becomes undefined. On the other hand, Method 2 can be simply applied to this function, as follows. Calculating the required sensitivities, the expression for the approximate change in f is

$$\Delta f \simeq \left((Ay^b z^{-c}) \cdot ax^{a-1}\right) \Delta x + \left((Ax^a z^{-c}) \cdot by^{b-1}\right) \Delta y$$
$$+ \left((Ax^a y^b) \cdot (-c)z^{-c-1}\right) \Delta z \tag{12.17}$$

where, as always, the sensitivity coefficients are calculated using the measured values of x, y, z and the known constants. Dividing by the given function f to get the relative change gives

$$\frac{\Delta f}{f} \simeq \frac{(Ay^b z^{-c}) \cdot ax^{a-1}}{Ax^a y^b z^{-c}} \Delta x + \frac{(Ax^a z^{-c}) \cdot by^{b-1}}{Ax^a y^b z^{-c}} \Delta y$$
$$+ \frac{(Ax^a y^b) \cdot (-c) z^{-c-1}}{Ax^a y^b z^{-c}} \Delta z$$
$$= a \frac{\Delta x}{x} + b \frac{\Delta y}{y} + (-c) \frac{\Delta z}{z} \qquad (12.18)$$

which is a linear combination of the relative uncertainties in the variables, with coefficients a, b, $-c$, which are the exponents of the variables in the term. Taking absolute values to get the worst-case result gives the approximate relative error as the simple expression shown:

$$\left|\frac{\Delta f}{f}\right|_{max} \simeq |a|\left|\frac{\Delta x}{x}\right| + |b|\left|\frac{\Delta y}{y}\right| + |-c|\left|\frac{\Delta z}{z}\right|. \qquad (12.19)$$

Example 11. Algebraic term: Example 2, mass of conical volume.

In contrast to the sensitivity constants for the recycled-glass example of Example 5, the relative sensitivities are simply the powers of the variables in the expression

$$m(\rho, h, d) = \frac{\pi}{12} \rho h d^2,$$

giving

$$\left|\frac{\Delta m}{m}\right|_{max} \simeq |1|\left|\frac{\Delta \rho}{\rho}\right| + |1|\left|\frac{\Delta h}{h}\right| + |2|\left|\frac{\Delta d}{d}\right|.$$

Thus, for relative changes, the computed quantity is most sensitive to the measured value of the diameter d, which is squared.

3 Method 3: Estimated uncertainty

The third method for calculating the uncertainty of f assumes a context that differs slightly from that of the first two methods. In particular, it may be possible to estimate the typical magnitude of random measurement errors rather than their exact range [1–3].

As before, the notation $f(x, y, z, \cdots)$ will represent the function to be calculated from inexact values $x = x_0 + \Delta x$, $y = y_0 + \Delta y$, $z = z_0 + \Delta z$, $\cdots$.

Random errors in its arguments will produce a random error in f. A positive change in f caused by error in one variable may be partially cancelled by a negative change caused by independent error in another, and the more variables there are, the less likely that the effects of their errors will combine in the same direction, as assumed in the first two methods. Rather, the method of this section is used to estimate the typical magnitude of the random error in f, instead of worst-case magnitude.

For the typical value of $|\Delta x|$, we shall use the standard deviation of x, also called the standard error, a statistical term [2, 3], to be discussed further in Chapter 13, defined in the current context as the square root of the average value of $(\Delta x)^2$, with a similar definition for other variables. We are required to find a typical $|\Delta f|$, which will be done by finding the standard deviation of f.

Assumptions:

1. The standard deviation of the value of the function is required, rather than the maximum deviation.
2. The errors in the variables are independent random quantities symmetrically distributed about a mean of zero, and the standard deviations of the variables are known.
3. The function is differentiable at the measured values of its arguments.

Under the above assumptions, the approximate change in the calculated value of f for changes in its arguments is given by Equation (12.4), where the formulas for the partial derivatives of f with respect to its arguments have been evaluated at the measured values to calculate the sensitivity constants $S_x, S_y, S_z \cdots$.

A simple derivation will be deferred to Section 3.1, but here the required result will be simply stated as follows:

$$\text{standard deviation of } f = \sqrt{S_x^2 |\Delta x|^2 + S_y^2 |\Delta y|^2 + S_z^2 |\Delta z|^2 + \cdots} \quad (12.20)$$

where the values of $|\Delta x|, |\Delta y|, \cdots$ inserted in the formula are the standard deviations of $x, y, \cdots$ respectively, and the positive square root is used.

Example 12. Estimated error in Example 1, resistance.

The measured voltage V in Example 1 will be assumed to have standard deviation equal to 0.5, rather than a range of exactly ± 0.5 as before. Similarly the measured current will be assumed to have standard deviation of 0.005, so that, using the sensitivity constants previously calculated in Example 4, the error in the calculated resistance is

$$\text{standard deviation of } R = +\sqrt{S_V^2 |\Delta V|^2 + S_I^2 |\Delta I|^2}$$
$$= +\sqrt{(0.448)^2 |0.5|^2 + (-23.5)^2 |0.005|^2}$$
$$= 0.25,$$

giving the result that the calculated resistance is $R = (52.5 \pm 0.25)\,\Omega$, where the uncertainty is a statistical value rather than a maximum value as it was in Examples 1 and 7.

3.1 Derivation of the estimated value

The origins of Equation (12.20) can be developed as follows. In Equation (12.4), repeated here,

$$\Delta f \simeq S_x \Delta x + S_y \Delta y + S_z \Delta z + \cdots, \qquad (12.4)$$

each of the quantities Δx, Δy, Δz, $\cdots$ is now assumed to be random, with zero mean value. To calculate the standard deviation of f, we shall square both sides and take the positive square root of the expected (mean) value. Squaring gives

$$(\Delta f)^2 \simeq S_x^2 (\Delta x)^2 + S_y^2 (\Delta y)^2 + S_z^2 (\Delta z)^2 + \cdots$$
$$+ 2 S_x S_y \Delta x \Delta y + 2 S_x S_z \Delta x \Delta z + 2 S_y S_z \Delta y \Delta z + \cdots. \quad (12.21)$$

In calculating the expected value, or mean, all products of the form $(\Delta x)^2$,

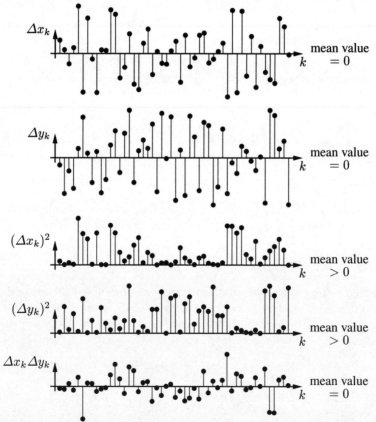

Fig. 12.3. Graphs of samples of zero-mean independent random variables Δx_k and Δy_k, $k = 1, \cdots$, showing that the mean of the squared quantities is positive, but the products $\Delta x_k \Delta y_k$ may be positive or negative, and have an expected (mean) value of zero.

$(\Delta y)^2, \cdots$ are non-negative, and therefore have non-negative mean value. However, because $\Delta x, \Delta y, \cdots$, are random, independent, and symmetrically distributed about zero, products of the form $\Delta x \Delta y, \Delta x \Delta z, \cdots$, may be either positive or negative, and will have expected value of zero, as illustrated in Figure 12.3. From the above reasoning, we conclude that

expected value of $(\Delta f)^2 \simeq$
$$\text{expected value of } \left(S_x^2 (\Delta x)^2 + S_y^2 (\Delta y)^2 + S_z^2 (\Delta z)^2 + \cdots\right)$$
$$= S_x^2 \times (\text{expected value of } (\Delta x)^2)$$
$$+ S_y^2 \times (\text{expected value of } (\Delta y)^2) + \cdots. \tag{12.22}$$

This is a standard result [2, 3], often called the "propagation of error formula." By the definition of the standard deviation of f, the positive square root of both sides of Equation (12.22) is now taken, giving Equation (12.20), as required.

4 Further study

1. Consider three measurements x, y, z, as listed below along with estimates of the error in each measurement:

    $$x = 6.31 \pm 0.04, \quad y = 9.23 \pm 0.01, \quad z = 16.3 \pm 0.5,$$

    and consider also the derived quantities d, e, f, g, shown below:

    $$d = x + 2y + 3z, \quad e = xy/z, \quad f = x^2 y z^{0.5}, \quad g = (x/z) + 3y$$

    (a) Compute the nominal values of d, e, f and g.
    (b) Using the exact-range method, compute the maximum and minimum values for d, e, f and g.
    (c) Using the linear approximation method, compute the error in each derived quantity.
    (d) Using Method 3, and assuming that the measurement uncertainties given above are independent estimated errors, compute the estimated error in each derived quantity.

2. A mass m hung by a string of length ℓ forms a pendulum that oscillates with a period T. The equation for the period of the pendulum can be derived easily and used to find the acceleration of gravity. From such an analysis, the formula for the period is $T = 2\pi \sqrt{\ell/g}$; rearranging, $g = 4\pi^2 \ell T^{-2}$.

    The string length is measured to be $\ell = (2.000 \pm 0.003)$ m, and the time for the pendulum to make 10 complete oscillations is $10T = (28.5 \pm 0.5)$ s.

    (a) Calculate the acceleration g of gravity.
    (b) Calculate the percent error in g using the three methods given in this chapter.

3. The air drag coefficient of an automobile typically is not measured directly, but is calculated from observations of the vehicle motion. Drag can be reduced by streamlining. With the automobile rolling in neutral, we observe coast-down times, which are the time intervals required to lose a given velocity. Two friction forces are involved: the rolling friction of tires, axles, and wheel bearings, which are nearly independent of the velocity, and the air drag friction, which is proportional to the square of the velocity. By making observations at two different velocities, the rolling friction can be eliminated, and the air drag friction can be isolated. Assume that for a vehicle of mass M, we measure the interval Δt_1 to coast from $v_1 + \frac{\Delta v}{2}$ to $v_1 - \frac{\Delta v}{2}$, and the interval Δt_2 to coast from $v_2 + \frac{\Delta v}{2}$ to $v_2 - \frac{\Delta v}{2}$. The difference between the total frictional forces at v_1 and v_2 gives the difference between the air drag forces alone, because the other frictional forces cancel out, and the following equation results:

$$M \frac{\Delta v}{\Delta t_1} - M \frac{\Delta v}{\Delta t_2} = A C_D k_1 k_2 (v_1^2 - v_2^2),$$

where the drag coefficient C_D can be solved in terms of the other quantities, and where:
M = vehicle mass in kilograms,
Δv = velocity decrement in kilometres per hour,
v_1, v_2 = mid-range velocities in kilometres per hour,
Δt_1, Δt_2 = coast-down time intervals in seconds,
$k_1 = 0.613$ kg/m^3 = a constant that is half the density of air at 15°C and 1 atm, proportional to air pressure divided by absolute temperature,
$k_2 = 1/3.6 = 0.277\,78$ = unit constant to convert kilometres per hour to metres per second,
A = the frontal projection of the maximum cross-sectional area of the vehicle, (square metres), approximately $0.83 \times$ height $\times$ width,

In coast-down tests on a certain automobile, the following are observed:
$v_1 = (55 \pm 1)$ km/h, $\quad \Delta t_1 = (27.7 \pm 0.2)$ s
$v_2 = (25 \pm 1)$ km/h, $\quad \Delta t_2 = (134.2 \pm 0.2)$ s
$\Delta v = (10 \pm 2)$ km/h

In comparison with the errors in the above dynamic observations, the quantities $M = 1200$ kg and $A = 2.21$ m^2 can be assumed to be error-free.

Calculate the drag coefficient C_D from the above data, and estimate the relative errors of the result using the three methods of this chapter.

4. The length x and width y of a plot of land are to be measured, with estimated errors Δx and Δy respectively.

(a) Sketch the rectangle, with measured dimensions x, y.

(b) On the sketch, superimpose the rectangle that corresponds to measurements $x+\Delta x$, $y+\Delta y$, assuming positive Δx, Δy.

(c) On the sketch, identify the change in area between the two measurements, as estimated by Method 2 of this chapter.

(d) On the sketch, identify the area that is the difference between the actual change in area and the estimated change in area.

Answers to problems:

1. (a) Using the nominal measurement values, the formulas evaluate to
   $$d = 73.7, \quad e = 3.57, \quad f = 1.48 \times 10^3, \quad g = 28.1.$$
   (b) The functions are all monotonic with respect to their arguments, so the extreme values are found at extreme values of the arguments as
   $$72.1 \leq d \leq 75.2 \qquad\qquad 3.44 \leq e \leq 3.71$$
   $$1.44 \times 10^3 \leq f \leq 1.53 \times 10^3 \quad 28.0 \leq g \leq 28.1$$
   (c) The approximate error of each computed quantity is given by the formula $|S_x||\Delta x| + |S_y||\Delta y| + |S_z||\Delta z|$ where, for each function, the formulas for the partial derivatives with respect to x, y, and z are evaluated at the measured values to give the sensitivity constants S_x, S_y, S_z. The symbol $|_0$ associated with a formula indicates that the formula is to be evaluated at the measured values of its arguments.
   $$d: S_x = \left.\frac{\partial d}{\partial x}\right|_0 = 1, \quad S_y = \left.\frac{\partial d}{\partial y}\right|_0 = 2, \quad S_z = \left.\frac{\partial d}{\partial z}\right|_0 = 3,$$
   $$|\Delta d| \simeq 1.56$$
   $$e: S_x = \left.yz^{-1}\right|_0 = 0.566, \quad S_y = \left.xz^{-1}\right|_0 = 0.387,$$
   $$S_z = \left.xy(-1)z^{-2}\right|_0 = -0.219, \quad |\Delta e| \simeq 0.136$$
   $$f: S_x = \left.2xyz^{1/2}\right|_0 = 470, \quad S_y = \left.x^2 z^{1/2}\right|_0 = 161,$$
   $$S_z = \left.x^2 y z^{-1/2}/2\right|_0 = 45.5, \quad |\Delta f| \simeq 43.2$$
   $$g: S_x = \left.z^{-1}\right|_0 = 0.0613, \quad S_y = 3,$$
   $$S_z = \left.x(-1)z^{-2}\right|_0 = -0.0237, \quad |\Delta g| \simeq 0.0443$$
   (d) The approximate error of each computed quantity is given by
   $$\sqrt{S_x^2 \Delta x^2 + S_y^2 \Delta y^2 + S_z^2 \Delta z^2}$$
   where the sensitivities for each function were calculated in the previous problem:
   $$|\Delta d| \simeq 1.50, \ |\Delta e| \simeq 0.112, \ |\Delta f| \simeq 29.6, \ |\Delta g| \simeq 0.0324.$$

2. (a) The acceleration of gravity $g = 9.72 \text{ m/s}^2$.
   (b) The relative errors are: (i) exact range: -3.56%, $+3.76\%$; (ii) linear approximation: $\pm 3.66\%$; (iii) estimated: $\pm 3.51\%$.

3. Solving for C_D and substituting in the formula gives $C_D = 0.380$. The relative errors in C_D, calculated by the three methods, are

(a) Exact range errors: $+29.9\,\%$, $-25.3\,\%$;
(b) Method 2: $\pm 27.6\,\%$.
(c) Method 3: $\pm 20.6\,\%$.

The sensitivities of the relative error in C_D to the relative errors in the measurements are: $\Delta v : 1$, $v_1 : 2.52$, $\Delta t_1 : 0.521$, $v_2 : 1.260$, $\Delta t_2 : 0.260$.

The results show that the error in Δv dominates the other sources of error, even though the error in the result is more sensitive to two of the others. In fact, we can see that the error in Δv needs to be reduced by a factor of 4 or 5 in order to achieve useful results. On the other hand, improving the time measurements would achieve nothing. The sensitivities, combined with the errors in the measurements, help us to understand how errors might arise in the final result.

4. The diagram below shows the measured area xy and the area $(x+\Delta x)\times(y+\Delta y)$ implied by the positive deviations Δx and Δy. The difference between the change in area and the linear estimate of the change in area is the rectangle $\Delta x \Delta y$, since the linear estimate is the quantity $x\Delta y + y\Delta x$.

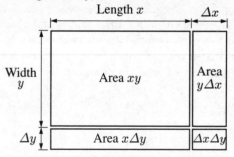

5 References

[1] N. C. Barford, *Experimental Measurements: Precision, Error and Truth*, John Wiley & Sons, New York, 2nd edition, 1985.

[2] R. V. Hogg and J. Ledolter, *Applied Statistics for Engineers and Physical Scientists*, Macmillan, New York, 1992.

[3] S. B. Vardeman, *Statistics for Engineering Problem Solving*, IEEE Press, New York, 1994.

Chapter 13

Statistics

A knowledge of statistics can guide the engineer in the collection, organization, analysis, and interpretation of measured data and is particularly important for quantifying and evaluating random measurement errors [1]. Systematic errors can be removed by calibration, or by correcting mistakes in measurement or computation. Statistics can be used to reduce the effect of random errors and to establish a level of confidence in measurements.

Chapters 11 and 12 discussed inexact numbers with a range of uncertainty or inexactness, which is specified. Often, uncertainty must be determined through experimentation and observation. For example, in a simple measurement, such as the length of a desk or table, random error will contribute to the uncertainty, or inexactness, of the measured length.

In this chapter, you will learn

- methods for calculating the central value of a set of measured data values, such as measuremets of length, voltage, temperature, or other basic quantities.

- methods for calculating the spread associated with a set of measured values,

- a method for calculating the relative standing of a measurement, for example, determining if your salary is in the lower quarter of comparable salaries; and

- how to present measured data using frequency distributions such as histograms.

1 Basic definitions

Some basic statistical terms to be used later are defined below:

Observation: An observation is simply another term for a measured value, such as a voltage reading from a voltmeter.

Parent population: The parent population for a measurement variable is the complete set of all possible observations for the variable. For a manufactured item such as a ball-bearing, its parent population is the set of all ball-bearing radii assuming that measurements of ball-bearing radii could be repeated indefinitely, yielding an infinitely large parent population. Statistics of the parent population are usually identified using Greek letters, such as μ and σ.

Parent distribution: The parent distribution describes the frequency of occurrence with which parent population measurements occur. If the set of all possible measured values corresponds to an interval on a line representing the real numbers, then the function is continuous. If the set corresponds to a sequence of points on the line, the function is said to be discrete. As discussed in detail in Chapter 14, the Gaussian distribution is an example of a continuous parent distribution that is commonly encountered when measurements are affected by independent sources of random variation.

Sample: A sample is a set of observations consisting of one or more measurements taken from the parent population. Equivalently, a sample is a subset of the parent population. In the ball-bearing example mentioned previously, the radius values actually measured for 10 ball-bearings is a sample set and is taken from the parent set that contains all possible radius measurements. If the parent population is infinitely large, then the possible number of different sets of samples is also infinitely large. Statistics of a sample are usually identified using English letters, such as m and s.

Sample distribution: A sample distribution is a function that shows the frequency with which individual observations within a given sample occur. A sample distribution is depicted in Figure 13.1, which uses data from Example 1 on page 154. Since the number of observations in a sample is finite, a sample distribution is always discrete.

Error: An error is the difference between a measured value of a variable and the variable's true value; it is the total effect of both systematic errors (bias) and random errors (imprecision), as discussed in Chapter 11. The true value is usually not known; otherwise, there would be no reason to perform a measurement.

Statistic: A statistic is a parameter that characterizes data in the presence of random uncertainties. The arithmetic mean, for example, provides one method to characterize the central value of examination grades.

1.1 Engineering applications of statistics

A knowledge of statistics is of value to engineers in many situations, of which the following four examples are typical.

True value of a single measurement: The elevation change between a reference point on a construction site and a construction stake elsewhere on the site is required. A true value exists, but an observation will probably be in error. In this case the engineer substitutes the best estimate of the true value along with an estimate of the range in which the true value is expected to be found. The value is written as described in Chapter 11, for example, as (7.15 ± 0.12) m.

Difference between measured and desired values: Quality-control tests of electrical resistors are conducted by measuring the resistance of sample resistors taken from the manufacturing line. Variations in the manufacturing process cause random variations in resistance. Monitoring the difference between desired and measured resistance allows a check that the manufactured values are within specified tolerance limits and also allows an estimate of the distribution of resistance values.

Central value and spread: Noise heard on radio signals is often random, due to imperfections in atmospheric transmission, thermal effects, and other causes. Electronic filtering in the radio receiver extracts an estimated value of the desired signal (the central value) in the presence of noise (spread). A radio capable of scanning frequencies and stopping on good signals uses information about the signal (central value) and noise (spread) to separate clear signals from noisy or weak signals. This information is often expressed as a signal-to-noise ratio.

Spread in measurements: The spread in a measurement may be required, for example, when determining the required overcapacity for an electricity utility. One objective of an electricity utility is to ensure that electricity can be delivered when demand rises, but a second objective is to minimize costs by minimizing total power generation capacity. Brownouts, or worse still blackouts, can have serious negative economic and societal impacts, such as a loss in manufacturing capability, or a loss of safe lighting. The statistical spread in electricity demand coupled with modelling of future electrical demand assist the electrical utility in maintaining an optimum overcapacity.

1.2 Descriptive and inferential statistics

The discipline of *statistics* involves the description of data and inferences made from the properties of data [2]. The goal of *descriptive statistics* is to describe or characterize data in simple ways, often for the purpose of comparison with other descriptive statistics. One goal of *inferential statistics,* sometimes called "sampling" statistics, is to estimate the properties of a parent population from a sample of that population for the purpose of prediction. Other goals of inferential statistics include hypothesis testing, risk assessment, and decision-making.

Descriptive statistics provide statements of fact about a sample, whereas inferential statistics provide estimates concerning the parent population of the sample. Table 13.1 contains examples of description and inference.

The remainder of this chapter discusses some elementary descriptive statistics, while Chapter 14 introduces inferential statistics as well as the slightly less elementary descriptive statistics of correlation.

Three classes of descriptive statistics are discussed in this chapter: measures of central location, spread, and relative standing. The measures of central location are the median, mode, and mean; spread is described by the range, variance, and

Table 13.1. Examples of description and inference. Samples are always described, but the properties of a parent population may be inferred from those of a sample.

	Example	Information provided
1	A class is tested on their knowledge of thermodynamics.	The class grade average provides a description of the entire class (the parent population). Inference is not necessary.
2	A hockey player is given a contract on the basis of ability to score goals.	Inference is required, from the number of goals scored in past games to future performance. Past goals are the sample; past and future goals are the parent population.

standard deviation; and relative standing is described in terms of quartiles, deciles, and percentiles.

Example 1. A sample: Student grades on a 20-point scale.

A language school assigns test marks as integers from 0 to 20. The school has kept track of the marks obtained on every language test they have ever administered. These grades are given in Table 13.2. For this sample set of all past test marks, the corresponding sample distribution is shown in Figure 13.1 and represents the best estimate of the parent distribution of grades. It is not the actual parent distribution, since the actual parent population is the set of all possible observations, including future observations. The distribution is a bell-shaped function with a

Table 13.2. Mark occurrences and relative frequencies for past language tests.

Mark	0	1	2	3	4	5	6
Occurrences	1	0	2	7	28	86	163
Rel. freq.	0.000	0.000	0.000	0.000	0.000	0.001	0.002
Mark	7	8	9	10	11	12	13
Occurrences	367	889	1845	3311	5392	7944	10 472
Rel. freq.	0.004	0.009	0.018	0.033	0.054	0.079	0.105
Mark	14	15	16	17	18	19	20
Occurrences	12 506	13 393	13 034	11 431	8869	6272	3856
Rel. freq.	0.125	0.134	0.130	0.114	0.089	0.063	0.039

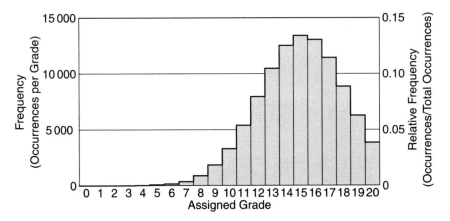

Fig. 13.1. Language test mark distribution for all past tests as given in Table 13.2. The bin size of this histogram is 1 and, using the relative frequency scale on the right, the area under the function is 1.

peak value at 15. The relative frequency of occurrences is calculated by dividing the number of occurrences of a given mark by the total number of observations. A relative frequency distribution is useful for distribution shape comparison; for example, if 110 people wrote the language test at the last sitting, the shape of the relative frequency grade distribution of this much smaller sample can be directly compared to Figure 13.1 because the areas under both curves would be the same.

Looking for and making independent checks of calculations is one method used by engineers to ensure the quality of their work. For example, the sum of relative frequencies given in Table 13.2 must, by definition, equal 1 when calculated exactly, and the sum of the given fractions is an independent check on the calculations of relative frequencies. The sum of the relative frequencies in Table 13.2 actually equals 0.999; the slight difference from 1 is because of rounding errors introduced in converting fractions to approximate decimal values. The difference between the actual sum and 1 is negligible.

In this special case, when the area under a relative frequency histogram equals 1, the histogram is said to have been normalized; consequently, the column height expressed as a relative frequency provides an estimate of the probability of observing a given grade value. Probability estimates is one of the statistical inferences discussed in Chapter 14.

2 Measures of central value

The mean, median, and mode are the three most common measures of the central location for a set of observations [3]. These are discussed and defined below.

Mean: Two means will be defined: the *sample mean* $\bar{x}$ and the parent *population mean* μ. The population mean has two variants, depending on whether the parent population has a finite or infinite number of possible observations. Only one variant of the sample mean exists, since a sample is always composed of a finite number of observations.

In the following, consider that a sample of n observations x_i, $i = 1, \cdots n$, has been taken from a parent population of N observations where $1 < n \leq N \leq \infty$. The sample mean is then defined as

$$\bar{x} = \frac{1}{n} \sum_{i=1}^{n} x_i. \tag{13.1}$$

If $N < \infty$, then the population mean is defined as

$$\mu = \frac{1}{N} \sum_{i=1}^{N} x_i; \tag{13.2}$$

otherwise, the population mean is defined as

$$\mu = \lim_{N \to \infty} \left(\frac{1}{N} \sum_{i=1}^{N} x_i \right). \tag{13.3}$$

The number of observations in the above equations must be 1 or greater; otherwise, the sums do not exist and the mean is undefined.

The majority of engineering statistical work involves sample descriptions and inferences, rather than parent population values, so informal use of "the mean" is implicitly taken to be the sample mean. In this book, the mean will refer to either the sample or the population. You must explicitly identify the meaning in documents where the context does not clearly identify the mean being used.

The sample mean is an important measure of central location because it can be proven to be the best estimate of the (unknown) population mean [4]. In this context, "best" is defined as the value that minimizes the mean squared error, as demonstrated in Section 2.1 of Chapter 14. Nevertheless, the sample mean is not always best for describing the central location; sometimes the median or mode, discussed next, is better suited.

Median: The median is the middle value when data is ordered (sorted) by magnitude. That is, half the observations will be less than the median and half will be greater. In terms of probability, this means that the probability is 0.5 (or 50 %) that any measurement x_i will be smaller than the median. Similarly, the probability is 0.5 that any measurement will be larger than the median.

If there are n observations, the *median* is defined as follows:

> When n is odd, the median is the value of the central observation such that $(n-1)/2$ observations have values below (less than or equal to) the median, and $(n-1)/2$ observations have values above (greater than or equal to) the median.

When n is even, the median is the mid-point value between the values of two most central observations. There will be $n/2$ observations above the calculated median, and $n/2$ observations below it. The two observations closest to the division between the upper and lower $n/2$ observations are the two most central observations. When n is even, the median does not equal an actual observation value when the two most central observation values are not equal.

Consider a family that has five children of ages 8, 9, *12*, 13, and 19, where the median age 12 has been written in italic type. If the 19-year-old child is not living at home, then the median age of the children still at home is taken from the sample 8, *9*, *12*, and 13, where the two central observations are italicized, and the mid-point between them gives a median age of 10.5.

Generally, the median should be used when a central value is sought for a distribution that is significantly asymmetric, as shown in Example 2. An asymmetric distribution is said to be *skewed*, as discussed in more detail in Section 2.1. It is generally clear from context when a sample or parent population median is being discussed.

Example 2. Salaries in a small company.

Consider the salary distribution in a company, as shown in Figure 13.2. The two outliers to the right of the distribution are the salaries of the two owners of the company. To plot the histogram, salaries have been placed in groups, or bins, $5 000 wide. For an employee trying to judge his or her salary level relative to others, the median is most meaningful, since only 2 of the 14 people make more than the mean, and these 2 make more than three times the mean. If the

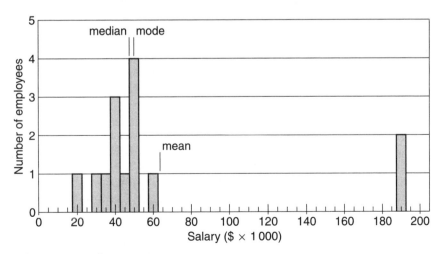

Fig. 13.2. The distribution of salaries in a small company.

employee wishes to know how many people earn more than she or he does, then relative standing statistics, as described in Section 4, become relevant.

Mode: The *mode* of a distribution is the value that is observed with the greatest frequency. This value is usually the simplest central value to identify.

For quantitative data that can be ordered, the mode is used when a very quick estimate of central value is needed. It is also used specifically to identify the typical (most common) observation value.

For quantitative data that cannot be ordered, such as the number of different types of cars rolling off assembly lines daily or enrollments in different subject-matter courses, the mode is really the only central location measure that can be used.

A distribution with one peak value is said to be unimodal; it is multimodal if there exist well-separated peaks. A single peak, and hence single mode, usually occurs when the observations are subject to random influences, and large random errors are less likely than smaller errors. When two or more well-separated peaks occur in a sample distribution obtained from a large number of observations, there are usually systematic influences on the observations.

As an example of a mode, for the bins of width $5 000 in Figure 13.2, the mode is $50 000.

In summary, the mean can be informally thought of as the "balance point" of a distribution, the median cuts the distribution in half, and the mode identifies the most frequent or most probable observation [3].

2.1 Mean, median, and mode of skewed distributions

The mean, median, and mode of a parent population are equal when observations greater than or less than the central value are equally likely. When small errors are more likely to occur than large errors, a unimodal distribution results. For a sample taken from a symmetric unimodal parent population, the mean, median, and mode of the sample distribution are typically equal or nearly equal. The mean, median, and mode are not equal for skewed (asymmetric) distributions.

A distribution is said to have a negative skew when the distribution has a tail that tends toward lower values and a central value that tends toward higher values, as illustrated in Figure 13.3. For a continuous, negatively skewed distribution that monotonically decreases from the mode, the order of the central values from smallest to largest is from mean, to median, to mode as shown in Figure 13.3. Student examination grades are often negatively skewed, since the majority of students are expected to receive a grade above 50 %, so that the tail extending to 0 % is longer than the tail to 100 %.

A distribution is said to have a positive skew when the distribution has a tail that tends toward higher values and a central value that tends toward lower

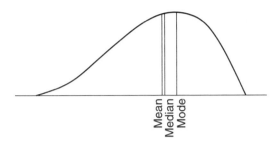

Fig. 13.3. A negatively skewed distribution.

values. For a continuous, positively skewed distribution that monotonically decreases from the mode, the order of the central values from smallest to largest is from mode, to median, to mean. Salaries are often positively skewed.

The skew of a distribution may be measured by the ratio of the mean to the median. A ratio greater than one corresponds to a positively skewed distribution, while a ratio less than one corresponds to a negatively skewed distribution.

3 Measures of spread

The range, variance, and standard deviation are the three commonly used measures of spread for a set of observations. Section 4.1 defines other measures—the interquartile and semi-interquartile ranges—as measures of spread.

Range: The *range* is simply the difference between the values of the largest and smallest observations.

Because it is so easy to determine, the range is frequently used in many simple tests, such as for quality control tests on a manufacturing line. Unfortunately, a major weakness of the range is its strong sensitivity to changes in data extremes due to its insensitivity to distribution shape. Therefore, the variance and standard deviation, described below, are preferred for more complex analyses, especially for the analyses of observations from symmetric or approximately symmetric distributions.

Variance: The spread in a distribution is related to the deviations of individual observations from the central location. From this perspective, one might propose a measure of spread that incorporates information about all observations by averaging the deviations $x_i - \mu$ between the observation value x_i and central location μ. However, the average of deviations for a symmetric distribution is zero due to the positive and negative nature of the deviations. One solution to the symmetry problem is to average the absolute value of the deviations $|x_i - \mu|$. Unfortunately, functions involving absolute values are cumbersome. The key to solving the symmetry problem is to find a measure of spread that depends on deviation magnitudes but not their sign and is easy to work with. Such a function is found by averaging the square of the deviations, and this function or measure of spread is the basis for defining the population variance.

Let N be the number of elements in a population. Then the *population variance* is defined as

$$\sigma^2 = \frac{1}{N} \sum_{i=1}^{N} (x_i - \mu)^2, \tag{13.4}$$

when $N < \infty$; otherwise,

$$\sigma^2 = \lim_{N \to \infty} \left(\frac{1}{N} \sum_{i=1}^{N} (x_i - \mu)^2 \right). \tag{13.5}$$

In the above definition, the true mean value μ is used, as well as the total population size N, which is often infinite. In most cases μ is unknown; hence σ^2 is unknown.

Using the sample mean $\bar{x}$ as a best estimate of the population mean μ naturally leads to the *mean-square deviation* (MSD) as a measure of sample spread. For a sample size n, the MSD then becomes a reasonable measure of spread to propose and define as follows:

$$\text{MSD} = \frac{1}{n} \sum_{i=1}^{n} (x_i - \bar{x})^2. \tag{13.6}$$

The MSD gives the spread (dispersion) about the sample mean for a given sample, but for $n < N$, the MSD can be shown to introduce a systematic error. Proving that this bias exists is beyond the scope of this discussion.

Given only the sample data, an estimate of the population variance can be found. This estimate is the *sample variance* and is defined as

$$s^2 = \frac{1}{n-1} \sum_{i=1}^{n} (x_i - \bar{x})^2. \tag{13.7}$$

The sample variance is the best estimate of the population variance when the population mean must be estimated using the same sample data, as is generally the case. The MSD is the best estimate when the population mean is independently known from theory and can be the better estimate when the population mean is independently known from a completely independent sample. For example, a sample length measurement, made using a ruler, is completely independent of a sample made using a laser device.

Standard deviation: The variance is an acceptable measure of the spread of a distribution. However, the units of the variance are the square of the units of the measurement. This prevents direct comparison of the variance with the mean. This undesirable situation is solved by taking the square root of the variance; therefore, the *sample standard deviation* s is given by $s = \sqrt{s^2}$, and the *population standard deviation* σ is given by $\sigma = \sqrt{\sigma^2}$. Generally, the standard deviation is the preferred measure of distribution spread.

4 Measures of relative standing

Quartiles: When observations can be ordered by magnitude, the quartile values divide the observations into four groups of equal number. At the upper end of the first quarter from the bottom is the lower quartile (LQ), at the upper end of the second quarter is the second quartile or median, and at the upper end of the third quarter is the upper quartile (UQ).

Deciles: Decile values divide an ordered set of observations into 10 groups of equal number. Of particular interest are the upper end of the first decile, which is referred to as the lower decile (LD), and the upper end of the ninth decile, which is referred to as the upper decile (UD).

Percentiles: Percentile values divide a set of observations into 100 groups of equal number.

The median, quartiles, deciles, and percentiles are useful for describing skewed distributions and are easy to determine. However, there must be a sufficient number of observations to make their determination meaningful. For example, if there are only three observations in the sample, it is meaningless to discuss observations below the lower decile because a sample of 3 cannot be divided into 10 groups of equal size.

4.1 Measures of relative standing spread

The interquartile range (IQR) can be used as a measure of relative standing spread. The IQR range measures the range between the lower and upper quartiles and is defined as

$$IQR = UQ - LQ. \tag{13.8}$$

By definition, 50 % of the observations are between the upper and lower quartiles. The interquartile range spans the median. If a measure of relative standing spread measured as a distance from the median is desired, then the semi-interquartile range (SIQR) is useful and is defined as

$$SIQR = (UQ - LQ)/2 = IQR/2. \tag{13.9}$$

The IQR and SIQR are functions of the quartiles, which, like the median, are independent of the magnitudes of the extremes of the distribution.

5 Presentation of measured data

The graphical presentation of data is a very important tool that can provide insight into data, ease its understanding, and assist in its interpretation. This section presents one type of graphic used to display distributions: the *histogram,* also known as a frequency diagram and, in some computer graphics software, as a "bar graph." To illustrate a histogram, data from a simple metal-fatigue experiment is used. The experiment is described in Example 3.

Example 3. A metal-fatigue experiment.

The objective of this experiment is to determine the fatigue properties of steel paper clips. The data is summarized in Table 13.3. The procedure followed by a group of first-year engineering students was to bend paper clips and then fully reverse the bend in a systematic way, repeating until the paper clip fractured. Each time a paper clip was bent from its starting position and fully back without breaking counted as one bend. The students counted the number of successfully completed bends for 662 paper clips until they broke.

Table 13.3. Observations from 662 paper clips fatigue-tested to failure.

Bends to failure	Observed frequency	Bends to failure	Observed frequency
0	0	11	65
1	5	12	54
2	1	13	24
3	7	14	29
4	7	15	10
5	26	16	8
6	58	17	11
7	77	18	5
8	95	19	3
9	89	20	4
10	83	21	1

Histogram: The histogram, or frequency diagram, illustrated in Figure 13.4, shows how the collection of 662 paper clips behaved in the simple metal-fatigue experiment of Example 3. The data is grouped (binned) by the number of full bend cycles a paper clip experienced before breaking. The left-hand y-axis of Figure 13.4 depicts the number of paper clips that failed per bin, that is, the frequency of failure; the right-hand y-axis depicts the number of paper clips that failed per bin relative to the total number of paper clips that failed, as a percentage. Regardless of the ordinate scale used, the shape of the distribution remains the same. The relative frequency scale is useful when different distribution shapes are to be compared. The frequency scale is useful when information about the actual number of observations in a particular abscissa (x-axis) bin is needed. Showing both scales is not necessary, but this is often done when it is not known which scale may be most useful to someone in the future wishing to interpret the data's distribution.

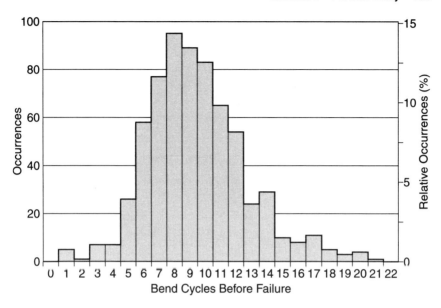

Fig. 13.4. A histogram of the metal-fatigue characteristics of steel paper clips from the experiment of Example 3.

5.1 Histogram interval selection

Suppose that the salary data presented in Figure 13.2 were organized into groups that spanned $1, instead of $5 000, such that, given the small sample size, no two people are placed in the same group. In this case, the small sample size does not lend itself to a histogram presentation. In addition, had only five observations been made in Example 3, a histogram would not have been useful. When the interval is too narrow, gaps (empty intervals) appear, and the histogram is a poor representation of the distribution. When the bin size is too wide, the histogram loses detail. Therefore, group data into intervals to make the corresponding histogram visually meaningful, as follows.

Histogram rule of thumb: include, as a minimum, 7 to 9 intervals over the entire data range, from the smallest to the largest observation value, being careful that the histogram bars near the central location contain at least 5 observations.

6 Further study

1. Find the mean, median, mode, range, variance, standard deviation, and semi-interquartile range for the following set of 17 grades for an examination marked out of 100:

25, 55, 60, 61, 63, 70, 72, 72, 73, 74, 74, 74, 76, 78, 82, 85, 90

2. Find the mean, median, mode, range, variance, standard deviation, semi-interquartile range, and semi-interdecile range for the language examination marks summarized in Table 13.2.

3. For the paper clip fatigue to failure data given in Table 13.3, calculate the mean, variance, and standard deviation using a computer spreadsheet program.

Answers to problems:

1. $\bar{x} = 69.6\,\%$, median $= 73\,\%$, mode $= 74\,\%$, range $= 65\,\%$, $s^2 = 213$, $s = 14.6$, SIQR $= 15$

2. *Hint:* The sum of the observation values in a given group (bin) is equal to the number of observations in that group multiplied by the observation value of that group.

3. *Hint:* Place the data in columns starting with the number of cycle bends to failure x_i, followed by the observation frequency n_i, the product $x_i y n_i$, and the product $x_i^2 ni$. Sum the last three columns, and use these sums to calculate the mean, variance, and standard deviation.

7 References

[1] J. S. Bendat and A. G. Piersol, *Random Data Analysis and Measurement Procedures*, John Wiley & Sons, New York, 1986.

[2] J. D. Petrucelli, B. Nandram, and M. Chen, *Applied Statistics for Engineers and Scientists*, Prentice Hall, Upper Saddle River, NJ, 1999.

[3] J. L. Phillips, *How to Think About Statistics*, W. H. Freeman and Company, New York, 6th edition, 2000.

[4] P. R. Bevington and D. K. Robinson, *Data Reduction and Error Analysis for the Physical Sciences*, McGraw-Hill, New York, 2nd edition, 1992.

Chapter 14

Gaussian law of errors

One major strength of the field of statistics is its ability to assist engineers in making conclusions and predictions. For example, consumers expect their automobile break pads to last an expected number of kilometres and to wear in a predictable fashion that ensures driver and passenger safety. To meet these stringent and critical demands, an engineer makes predictions of means, uncertainties, and risks on the basis of inferences made from a statistical analysis of break pad experiments and theory. Inferential statistics is the focus of this chapter and naturally leads to the concept of probability and the introduction of probability distributions. In the study of experimental observations and their associated random errors, the *Gaussian* (or *normal*) probability distribution is well accepted as being relevant in most circumstances as a consequence of the Gaussian law of errors. This contrasts with Chapter 13, where the focus of the statistics discussed was to describe data, not to make inferences from the data.

In this chapter, you will learn

- about the universality of the Gaussian law of errors in the statistical analysis of measurements and the conditions that must apply if the Gaussian law of errors is appropriate,
- about the relationship between probabilities, confidence intervals, and the Gaussian distribution,
- about the inference of a best estimate of population mean, a best fit line to a set of data, and when an observation can be rejected, and
- about the determination and interpretation of correlation coefficients as a descriptive statistic to measure the relationship between two sets of observations or between observation and theory.

1 The Gaussian law of errors

The Gaussian law of errors states that if measurement errors are the result of many uncontrolled random events, then the shape of the distribution describing the measurement is Gaussian. The Gaussian (or normal) probability distribution function is shown in Figure 14.1 and is defined by the following equation:

$$P(x) = \frac{1}{\sigma\sqrt{2\pi}} e^{-\frac{(x-\mu)^2}{2\sigma^2}}, \qquad (14.1)$$

where μ is the mean and σ is the standard deviation of the Gaussian distribution. The function given by Equation (14.1) is also known as a probability density function because the area under the curve from $-\infty$ to ∞ is 1; that is, the probability of an observation being within the range of all possible distribution observations is 100 %.

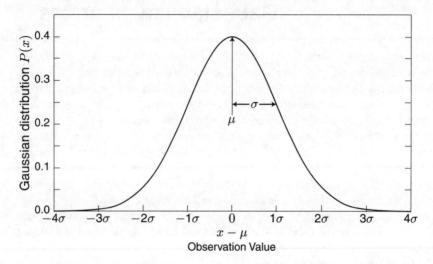

Fig. 14.1. The Gaussian (or normal) distribution with mean μ and standard deviation σ.

In 1809, Gauss used the normal distribution as a model for the errors in astronomical observations, resulting in its early description as the "law of errors." Since then, observations such as test scores, height, and many others have been observed to follow the Gaussian distribution. The normal distribution in statistics and the Gaussian distribution in engineering and the pure sciences are the same. In the special case when $\mu = 0$ and $\sigma = 1$, the distribution is called the *standard normal* or *standard Gaussian* distribution. Social scientists often refer to the Gaussian distribution as the "bell curve."

Example 1. Coin flipping follows the Gaussian law of errors.

For random flips of a coin, the probability of heads is just as likely as tails; hence, the probability distribution for a single coin flip is uniform, as shown in Figure 14.2. If the measurement of interest is the number of heads in a sample of n coin tosses, then the distributions shown in the figure for $n = 1, 2, 3,$ and 20 result. These are examples of a distribution said to be binomial. As the number of coin tosses n in a given sample increases, the distribution for the number of occurrences of heads observed approaches a Gaussian distribution. The emergence of a Gaussian distribution as n increases is a demonstration of the *central*

Section 1   The Gaussian law of errors   167

limit theorem, which states that the sum of a large number of independent, identically distributed observations will approach a Gaussian distribution regardless of the underlying distribution. The sum, in this case, is the number of heads, and the underlying distribution is shown in Figure 14.2. For large n, the resulting Gaussian distribution describes the distribution for the number of heads observed in a single measurement consisting of n coin tosses; this is a demonstration of the Gaussian law of errors in that many such measurements are expected to be Gaussian distributed.

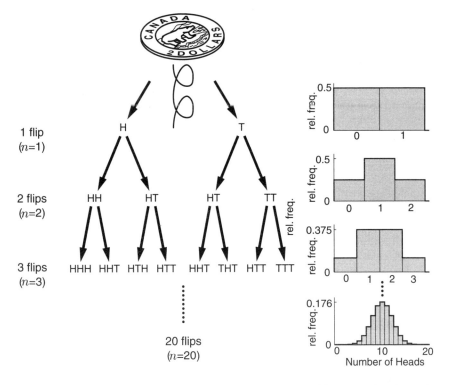

Fig. 14.2. The number of occurrences of heads observed in a sample of n coin flips demonstrates the central limit theorem through the emergence of a Gaussian distribution as n approaches infinity.

Example 2.   Noise in voltage measurements.

Many errors in engineering measurements and fluctuations in many other samples of observed data result from numerous, often very small, independent, random perturbations. A good example is "noise," caused by thermal electron motion, in low-voltage measurements. Regardless of the behaviour of these perturbations, the net effect is well described by the Gaussian law of errors, such that voltage-measurement errors from electron motion will be well described by the Gaussian probability distribution given in Equation (14.1).

The usefulness of the Gaussian law of errors is that it is often a reasonable model when large errors are less likely than small errors, it is of a fairly simple analytic form, and that much experimentation and theory demonstrate it to be the most likely distribution to occur.

Caution: The Gaussian law of errors does not always apply as a model for randomness. The frequent occurrence of Gaussian-like error distributions for samples taken from parent populations of unknown shape makes the Gaussian distribution a preferred choice for the error distribution. However, proofs that the Gaussian law applies exist only in special cases. A Gaussian error distribution is assumed when better information about the error distribution is unavailable, and there is no evidence that the conditions outlined in Section 1.1 below do not hold. As Lippman [1] states, "Everybody believes in the Gaussian law of errors: the experimenters because they think it can be proved by mathematics; and the mathematicians because they believe it has been established by observation."

1.1 Conditions for the Gaussian law of errors

The following five conditions should apply; otherwise, the applicability of the Gaussian law of errors may be suspect:

1. *Independent measurements:* Measurements (observations) are independent and equally reliable. The implication is that the measurement errors are random and independent.

2. *Symmetry:* Positive and negative errors of the same magnitude are equally likely; that is, errors are distributed symmetrically.

3. *Small errors are most probable:* Large errors are less likely than small errors. That is, the probability of an error of a given magnitude decreases monotonically with its magnitude.

4. *Mode equals mean:* This condition, that the mode equals the mean, follows from Conditions 2 and 3 above but is identified separately because of its usefulness in the absence of visualizing the error distribution.

5. *No cusp:* The derivative of the error distribution is continuous at the mode; hence, the derivative has a value of zero at the mode. That is, the error distribution and its slope vary continuously and smoothly everywhere.

The above five conditions are sufficient to determine the complete shape of the error distribution and were used by Gauss to do just that; the resulting distribution is given by Equation (14.1), the Gaussian distribution. Practically, this means that when the above-listed conditions exist, we are justified in assuming that the

Gaussian law of errors applies. Conversely, when we use Gaussian statistics, we are implicitly assuming that these conditions exist. Nevertheless, the Gaussian law of errors has shown itself to be quite robust and an excellent approximation, even when these five conditions are not precisely met. For example, a Gaussian distribution often provides a good description of the grade distribution observed on an examination even though grade distributions are, in general, asymmetric with a negative skew, as discussed in Section 2.1 of Chapter 13.

2 The Gaussian error distribution

The Gaussian probability density function, previously defined by Equation (14.1), shown in Figure 14.1, and redrawn in Figure 14.3 for the purposes of the following discussion, is also called a *Gaussian* (or *normal*) *error distribution* function.

The probability of an observation of a Gaussian-distributed variable falling within different intervals, as illustrated in Figure 14.3, is also given in Table 14.1.

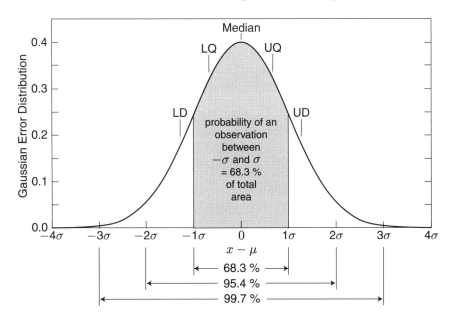

Fig. 14.3. The Gaussian error distribution function, also known as the Gaussian probability density function. The percentage area under the curve for selected intervals is shown.

Since the main purpose of the Gaussian distribution is to serve as a parent distribution model from which information about random fluctuations in a data sample can be inferred, it is expected that comparisons between Gaussian distributions and sample distributions will need to be analysed to confirm the validity

Chapter 14 Gaussian law of errors

Table 14.1. Standard Gaussian distribution function interval probabilities.

Interval			Probability of an observation falling in the interval (%)
-0.67σ	$\leq z \leq$	0.67σ	50
$-\sigma$	$\leq z \leq$	σ	68.27
-1.96σ	$\leq z \leq$	1.96σ	95
-2σ	$\leq z \leq$	2σ	95.45
-2.58σ	$\leq z \leq$	2.58σ	99
-3σ	$\leq z \leq$	3σ	99.73

of the Gaussian distribution model. As discussed in Chapter 13, the relative frequency diagram is appropriate for comparing the shapes of distributions and is obtained by normalizing the distribution function by the number of observations. The relative frequency diagram obtained from an experiment provides an estimate of the expected number of observations at a given observation value x_i for a discrete distribution, or between two observation values x_i and x_{i+1} for a continuous distribution.

The area under the normalized Gaussian error distribution is 1, so that the probability of an observation being in its domain is 1, or 100 %. Correspondingly, the area under the curve over the interval x_i to x_{i+1} equals the probability of an observation value occurring in this interval. Now imagine dividing the Gaussian error distribution into an infinite number of histogram bins of infinitely small width dx. Then the bin at the point x has height $P(x)$ and infinitely small area $P(x)\,dx$. The area $P(x)\,dx$ is equal to the probability of an observation in the interval x to $x+dx$. This can then be integrated from x_i to x_{i+1} to determine the probability of an observation in this interval. The Gaussian error function contains two parameters, μ and σ. Before integrating, it can be simplified using the variable

$$z = \frac{x - \mu}{\sigma}. \tag{14.2}$$

Equation (14.2) implies that $x = \mu + z\sigma$ and that $dx = dz$. The *standard Gaussian* (or *standard normal*) distribution is then given by

$$Y(z) = P(\mu + \sigma z) = \frac{1}{\sqrt{2\pi}} e^{-\frac{z^2}{2}}, \tag{14.3}$$

where the mean of z is zero, and its standard deviation is 1. The mode and median are also zero, so the standard Gaussian distribution is symmetrical about the origin, as shown in Figure 14.3.

If the population mean and standard deviation are known, then the probability of an observation in a given range, or interval, x_1 to x_2 can be determined as follows:

$$\text{Probability that } x_1 < x < x_2 = \int_{x_1}^{x_2} P(x)\,dx = \int_{z_1}^{z_2} Y(z)\,dz$$

$$= \int_{-\infty}^{z_2} Y(z)\,dz - \int_{-\infty}^{z_2} Y(z)\,dz, \quad (14.4)$$

where the standard Gaussian distribution integral

$$\int_{-\infty}^{z} Y(z)\,dz \quad (14.5)$$

is tabulated in books such as [2, 3] and is equal to the standard Gaussian cumulative distribution function value at z.

Figure 14.4 shows the integral of $Y(z)$ from $-\infty$ to z. The figure shows that the probability of an occurrence of z less than 0, or, from (14.2), of $x < \mu$, is 0.5, and the probability of any value occurring is 1.

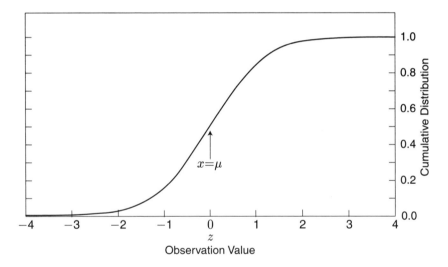

Fig. 14.4. The cumulative Gaussian error distribution function, which is the integral from $-\infty$ to x of $P(x)$. At x, the function shows the probability of occurrence of values $\leq x$.

If a random variable that is known to be Gaussian is measured, then μ and σ of the parent Gaussian distribution can be estimated from the observed data sample mean $\bar{x}$ and sample standard deviation s, respectively. Furthermore, it becomes possible to estimate the error in a single observation and in the sample mean. For example, for a single observation x_1, the precision to which this single observation measures the population mean, assuming no systematic errors, can be expressed with a *level of confidence* as follows: the population mean μ is known, with 68.3 % probability, to be in the range $x_1 \pm \sigma$, that is, with a 68.3 % level of confidence. It is common to use the intervals and confidence values in Table 14.1. The choice of confidence level is generally based on either convention or the demands of the problem being addressed by the measurement. In the single-observation case, the population standard deviation σ must be known beforehand or estimated from theory or other data.

When n measurements of a Gaussian-distributed variable are made, then the sample mean $\bar{x}$ provides the best estimate of μ. In this context, "best" is defined as the value that minimizes the mean square error (see Property one, Section 2.1 below). Furthermore, the uncertainty in the sample mean is represented by its standard deviation:

$$\sigma_{\bar{x}} = \frac{\sigma}{\sqrt{n}} \simeq \frac{s}{\sqrt{n}}, \tag{14.6}$$

where s is taken as an estimate of σ. Then one obtains

$$\mu = \bar{x} \pm \frac{ks}{\sqrt{n}}, \tag{14.7}$$

where k is the chosen confidence level, as in Table 14.1, that the actual population mean μ is within the range $\bar{x} - (ks/\sqrt{n})$ to $\bar{x} + (ks/\sqrt{n})$. For example, election polling to determine the percentage of votes that a political candidate will receive are commonly expressed as being "accurate to within 3 percentage points 19 times out of 20." If a particular candidate is expected to receive 35 % of the votes, then this statement is equivalent to saying that it is estimated that there is a 95 % probability that the actual percentage of votes received will be within the range $35(1 \pm 1.96\sigma)$ %. In effect, this 95 % confidence level in the 35 % measurement takes into account the limited size of the sample.

2.1 Properties of the sample mean

Property one: $\bar{x}$ *is a "best" estimate of* μ. Section 2 of Chapter 13 and Section 2 of this chapter state that the sample mean $\bar{x}$ is the best estimate of central value, in the sense that it minimizes the mean-square error. The proof requires that the sample mean-square error e^2, defined as follows, be minimized with respect to the central value estimate m:

$$e^2 = \sum_{i=1}^{n}(x_i - m)^2. \tag{14.8}$$

If the central-value estimate m that minimizes e^2 can be shown to be equal to the sample mean $\bar{x}$, then the proof will be complete. Using elementary calculus, e^2 can be differentiated with respect to m to yield

$$\frac{de^2}{dm} = \sum_{i=1}^{n} 2(x_i - m)(-1). \tag{14.9}$$

The first derivative is zero at a minimum or maximum of a function; therefore, at a minimum,

$$\sum_{i=1}^{n}(x_i - m) = 0 = \left(\sum_{i=1}^{n} x_i\right) - nm, \qquad (14.10)$$

which yields

$$m = \frac{1}{n}\sum_{i=1}^{n} x_i = \bar{x}. \qquad (14.11)$$

To complete the proof, the second derivative of e^2 with respect to m must be tested to ensure that $\bar{x}$ corresponds to a minimum and not a maximum. This can be accomplished by confirming that the second derivative of e^2 is positive.

Property two: *The sample mean $\bar{x}$ improves as an estimate of the population mean μ as the sample size n increases.* The details are not given here; this improvement is demonstrated in references such as [4]. As n increases, $\sigma_{\bar{x}}$ decreases as given in Equation (14.6). Therefore, the rate of improvement is proportional to $1/\sqrt{n}$.

Warning about statistics: Property two suggests that the error in the sample mean $\bar{x}$ of a set of measurements x_i can be reduced indefinitely by increasing the number of measurements, but this is not always true. Time and resources may inherently limit the number of measurements, systematic errors may exist, and non-statistical fluctuations such as occasional operator error may exist.

3 Validating the Gaussian law of errors

The histogram provides one graphical method to test data to see if the data follows a Gaussian distribution. This visual test, or validation, involves plotting a histogram based on the sampled data and then superimposing the Gaussian distribution to see how closely it fits.

3.1 Histogram of observed data

Consider the data provided in Table 13.3 from the paper clip metal-fatigue test. This data was plotted as a histogram in Figure 13.4. The objective now is to visualize how closely this histogram follows the shape of a corresponding estimated Gaussian parent distribution $\hat{P}(x)$.

The Gaussian distribution, defined in Equation (14.1), depends on two parameters: the population mean μ and the population standard deviation σ. The first step in obtaining $\hat{P}(x)$ is to calculate the sample mean $\bar{x}$ and standard deviation s. The histogram is a function giving occurrences with respect to x, and its area is the sum of the bin heights multiplied by bin width, or $n\Delta x$. The distribution

$\hat{P}(x)$ will be taken to have mean $\bar{x}$, standard deviation s, and to be the Gaussian distribution function scaled to have area $n\Delta x$:

$$\hat{P}(x) = \frac{n\Delta x}{s\sqrt{2\pi}} e^{-\frac{(x-\bar{x})^2}{2s^2}}. \tag{14.12}$$

Figure 14.5 shows $\hat{P}(x)$ superimposed on a histogram of the data. The sample mean, sample standard deviation, and scaling factor used in plotting the Gaussian distribution curve are 9.45 bends to failure, 3.19 bends to failure, and 662, respectively. Only discrete integer values for the abscissa variable (bend cycles to failure) exist, and $\Delta x = 1$, so the Gaussian distribution is an estimate of bin height only at these integer values. If the bin size Δx were different, then the scaling factor $n\Delta x$ would change, but the smooth curve would still provide one estimation point per bin as in the figure.

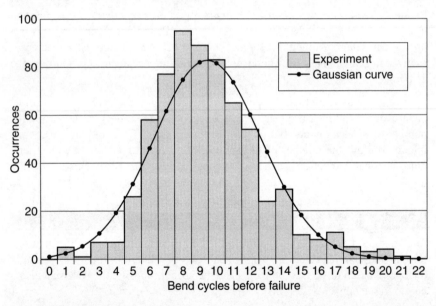

Fig. 14.5. Histogram of metal-fatigue test data from Example 3 with corresponding estimated Gaussian distribution $\hat{P}(x)$ superimposed for comparison. The estimated values are defined only at the dots, owing to the discrete nature of the histogram data.

Choice of histogram bin interval: The bin size Δx influences the appearance of the histogram and may affect your ability to visually validate the appropriateness of accepting the Gaussian law of errors for a given set of measurements. Too small an interval causes some intervals to be empty; too large an interval loses the bell shape. One way to solve this dilemma is to draw a histogram with non-uniform bin widths, with identical expected numbers of observations per bin, based on the assumed Gaussian parent distribution. In effect, each bin interval becomes "equally likely." There are then statistical tests, such as the chi-square (χ^2) test [2, 4], that can be used to determine if an observed deviation from exact uniformity is reasonable, that is, if a deviation from a constant number of observations in each bin interval is reasonable. Section 5.1 below provides a useful rule of thumb for selecting the histogram width interval in order to make a histogram visually meaningful.

4 The best straight line

The numerical method for obtaining the best straight line to fit a set of observed data points is called *linear regression,* which provides a mathematical alternative to drawing the best straight line through plotted data "by eye." The line is "best" in the sense of minimizing the square of the deviations between the straight line $y(x)$ and observation pairs x_i, y_i for $i = 1, \cdots n$. The equation of a straight line is

$$y(x) = ax + b, \tag{14.13}$$

where a is the slope, and b is the y-axis intercept. (The slope parameter is written as a in this chapter rather than m as in Chapter 9 because we have used m for the central value estimate.) Given n observation pairs, the linear regression method seeks to minimize the quantity chi-square parameter χ^2, which is the sum of the squared deviations, defined as

$$\chi^2 = \sum_{i=1}^{n}(y_i - y(x_i))^2 = \sum_{i=1}^{n}(y_i - ax_i - b)^2, \tag{14.14}$$

assuming that the errors $y_i - ax_i - b$ have equal standard deviation. To minimize this quantity, the unknown parameters a and b must be found. The solution turns out to require solving the following two simultaneous linear equations for unknowns a and b:

$$\left(\sum_{i=1}^{n} x_i\right) a + nb = \sum_{i=1}^{n} y_i \tag{14.15a}$$

$$\left(\sum_{i=1}^{n} x_i^2\right) a + \left(\sum_{i=1}^{n} x_i\right) b = \sum_{i=1}^{n} x_i y_i. \tag{14.15b}$$

In these equations, the right-hand sides and the quantities multiplying a and b are known constants, since n and the values of x_i and y_i are known. By any method, solving for a gives

$$a = \frac{n \sum(x_i y_i) - \left(\sum x_i\right)\left(\sum y_i\right)}{n \sum(x_i)^2 - \left(\sum x_i\right)^2}, \tag{14.16}$$

so that, from Equation (14.15a), b is

$$b = \frac{\sum y_i}{n} - a \frac{\sum x_i}{n}, \tag{14.17}$$

where the summation limits $i = 1$ and n have been omitted to simplify the notation. To derive Equations (14.15) from (14.14), use the fact that at a minimum, the partial derivatives (mentioned in Chapter 12) of a function of several variables are zero simultaneously. Thus, equating the partial derivatives $\frac{\partial \chi^2}{\partial b}$ and $\frac{\partial \chi^2}{\partial a}$ to zero yields Equations (14.15a) and (14.15b). This method has to be modified when the errors $y_i - ax_i - b$ have unequal standard deviation [4].

Warning about linear regression: Fitting a straight line to data is not always justified; the following items must be considered:

- *Unequal errors:* If the data points have unequal uncertainties, then the quantity χ^2, given by Equation (14.14), has to be modified. Figure 14.6 illustrates how equal data uncertainties become unequal uncertainties when the data are transformed using a non-linear function such as the logarithm.

- *Error in the abscissa variable:* The previous development assumes errors only in the quantities y_i, with the x_i known exactly. If the x_i quantities are also uncertain, a modified method must be used.

- *Extrapolation:* If the data have been measured over a limited range, generally it is dangerous to predict the value of $y(x)$ outside this range.

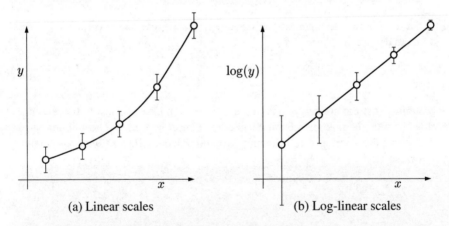

(a) Linear scales   (b) Log-linear scales

Fig. 14.6. The equal uncertainties of the data points in (a) become unequal under a nonlinear transformation, such as for the log-linear scales in (b).

4.1 Correlation coefficient

The parameters a and b in Equation (14.13) are easy to calculate from (14.16) and (14.17), even when it is blatantly clear that a straight line is not a good model for the data, either because of physical reasoning or because many different straight lines fit the data almost equally well. The *correlation coefficient* is a statistic that quantifies the degree to which the data fit a straight line.

The correlation coefficient r from a set of pairs x_i, y_i, $i = 1, \cdots n$, is calculated using the following equation, derived in references such as [3]:

$$r = \frac{\overline{xy} - \bar{x}\,\bar{y}}{\sqrt{\mathrm{MSD}(x)}\sqrt{\mathrm{MSD}(y)}}. \tag{14.18}$$

In this formula, $\mathrm{MSD}(x)$ and $\mathrm{MSD}(y)$ are the mean-square deviations of the x-values and y-values, respectively, defined in Equation (13.6), $\bar{x}$ and $\bar{y}$ are the respective sample means of the x-values and y-values, and $\overline{xy}$ is

$$\overline{xy} = \frac{\sum_{i=1}^{n} x_i y_i}{n}. \tag{14.19}$$

The values of the correlation coefficient r can be interpreted as follows:

If $r < 0$, then the data are said to be negatively correlated; if $r > 0$, they are positively correlated.

$r = 0$ indicates no linear dependency between y_i and x_i; the data are "uncorrelated."

$|r| = 1$ indicates that all data points are on the straight line; the relationship between y_i and x_i is perfectly linear.

$0 < |r| < 1$ is a quantitative measure of the degree of correlation.

Positive correlations imply a positive slope of the best straight line; similarly, negative correlation implies a negative slope. An example of positively correlated data is the increase in children's height with increasing age. An example of negatively correlated data is the decrease in rabbit population seen to accompany an increase in wolf population.

5 Rejection of an outlying point

Experimental data may contain "outlying" observations, that is, observations that seem highly improbable or unusual. Such outlying points generate concern about their validity because they may be a mistake or an improbable event that is not representative of the data set.

The first step in analysing outlying points is always to investigate the possible causes of an unusual observation. The unusual observation may signal an experimental process or natural phenomenon that is unexpected. If the observation clearly proves to be a mistake, then the observation should be rejected and the mistake corrected. If no mistake can be found, statistical tests are available to help decide if one is justified in rejecting the outlying observation. When a point is rejected for statistical reasons, it is not simply deleted from the data set; instead, it is recorded and then marked as "rejected," with the rejection criterion stated briefly.

Statistical rejection of an outlying observation is justified when it is sufficiently unusual that it adversely affects the calculations of the sample mean, sample variance, and sample standard deviation, *even if the observation is valid!* Validity implies no apparent cause for the outlier except an improbable random fluctuation or "bad luck."

Example 3. Rejecting the results from a coin flipping experiment.

A legal coin, with a head and tail, should show 50 % heads and 50 % tails if it is flipped a sufficient number of times. However, suppose that we flip a coin six times, with heads resulting each time. This is unusual, because it can be proven that the probability of this happening is only 1 event in 64, or 1.6 %. Therefore, although this result is valid, we would intuitively reject it. It would not alter our belief that a legal coin should show 50 % heads and 50 % tails if it is flipped. Of course, we might want to check the coin to see whether it has two heads. Similar logic is the basis for rejecting valid, but unusual, outliers.

5.1 Standard deviation test

How do we decide if an outlier is sufficiently unusual that it may be rejected as uncharacteristic of the data? The short answer is that it is a judgement that takes into account the type of questions the data is required to answer, knowledge of how the observations are obtained, the number of observations in the sample set, probability estimates based an assumed parent distribution, and other factors.

The *standard deviation test* is a common criterion for the rejection of outliers. It is based on the estimated probability of the occurrence of the outlier within the parent population, without regard to the sample size of the observation set. Following the definition of the standardized variable z introduced in Section 2, for suspected outlier x_i, let

$$z = \frac{x_i - \bar{x}}{s}. \tag{14.20}$$

In this formula, z is sometimes called the z-score, $\bar{x}$ is the sample mean, and s is the sample standard deviation. The standard deviation test for outliers is then simply as follows. An observation x_i is an outlier and may be rejected if

$$|z| > 3s. \tag{14.21}$$

By this criterion, any observation more than three standard deviations away from the mean is not considered representative of the population from which the rest of the sample is drawn and may be rejected. The discussion of the Gaussian error distribution in Section 2 showed that 99.73 % of Gaussian observations fall within $\pm 3\sigma$ of the mean μ. Therefore, the probability of obtaining an observation more than three standard deviations from the mean is 0.270 % for Gaussian random errors.

Caution: If the sample size is large and relevant or the errors do not follow a Gaussian distribution, then the standard deviation test may be inappropriate. To reject outliers requires that you understand your observations.

5.2 Application of outlier tests to skewed distributions

The above standard deviation test for outliers requires the errors to be Gaussian. An important class of observations must be treated with care, since the Gaussian distribution may not be applicable. This class includes data where only positive values are possible, which are therefore inherently skewed. Counting is an example, since the data consists of non-negative integers, such as persons, vehicles, or events. Other similar cases are class sizes, traffic statistics, and the number of earthquakes or floods. Other quantities are not integers but still cannot be negative, such as human height, weight, or lifespan. Measurements of these quantities do not satisfy the Gaussian symmetry condition for random errors discussed in Section 1.1.

To decide whether Gaussian statistics can be applied to cases of non-negative observations, calculate the ratio $\bar{x}/s$ of the sample mean to the sample standard deviation. For small values such as $\bar{x}/s < 5$, the distribution is badly skewed. However, as this ratio becomes larger, $\bar{x}/s > 20$, for example, the data distribution becomes bell-shaped, almost symmetrical about the mean, and approximately Gaussian. The standard deviation test can then be applied.

6 Further study

1. The four sets of data below represent coordinate pairs (x_i, y_i) for four distinctly different relationships. Using the equations in this chapter, calculate the slope a of the best straight line and the correlation factor r for each case. Make a quick plot in the x–y plane and comment on whether these values appear to be reasonable.

    Case 1: $\{(x_i, y_i)\} = \{(-2, 2)\ (-1, 1)\ (1, -1)\ (2, -2)\}$
    Case 2: $\{(x_i, y_i)\} = \{(1, 0)\ (0, 1)\ (-1, 0)\ (0, -1)\}$
    Case 3: $\{(x_i, y_i)\} = \{(-2, -2)\ (-1, -1)\ (1, 1)\ (2, 2)\}$
    Case 4: $\{(x_i, y_i)\} = \{(-2, 0)\ (-1, 0)\ (1, 0)\ (2, 0)\}$

2. A metal-fatigue experiment using paper clips was given as an example in Section 5 of Chapter 13. Using the standard deviation test for rejecting outlying observations, determine whether the extreme (upper or lower) observations should be rejected, as follows;

    (a) Find the $\pm 3s$ limit for rejecting outliers.

180 Chapter 14 Gaussian law of errors

(b) The $\pm 3s$ test is based on the Gaussian distribution for random errors. In this problem, the negative lower limit (-0.12 bends to failure) is clearly impossible and should alert us that something is wrong in the application of the outlier test to this problem. Which condition in Section 1.1 is violated in the metal-fatigue experiment data?

(c) What is the value of the $\bar{x}/s$ ratio? Can the standard deviation test be applied to outliers in this case? If so, how many outliers are there?

3. The cylinder pressure observed in an internal combustion engine varies from cycle to cycle due to random fluctuations in the combustion process. If 10 measurements of peak cylinder pressure are made, resulting in a mean peak cylinder pressure of combustion of (2.14 ± 0.02) MPa, how many additional measurements are needed to reduce the uncertainty to 0.01 MPa?

Answers to problems:

1. $(a, r) = (0, 0), (1, 1), (-1, -1) (0, 1)$
2. (a) $\bar{x} = 9.45$ bends to failure, $s = 3.19$ bends to failure, $-3s$ limit $= -0.12$ bends to failure, $+3s$ limit $= 19.02$ bends to failure.

   (b) The symmetry condition is violated.

   (c) $\bar{x}/s = 2.96$. Since $\bar{x}/s < 5$, the standard deviation test for outliers should not be applied. Another test may be considered to identify possible outliers, but such other tests are beyond the scope of this book.

3. A total of approximately 40 measurements is needed, assuming that only independent random errors affect the pressure measurements. Therefore, approximately 30 additional measurements are needed. The number of additional measurements needed is approximate because the uncertainty has been estimated using the sample standard deviation, which will vary from sample data set to sample data set.

7 References

[1] E. T. Whittaker and B. Robinson, "Normal frequency distribution," in *The Calculus of Observations: A Treatise on Numerical Mathematics*, pp. 164–208. Dover Press, New York, 4th edition, 1967.

[2] D. C. Montgomery and G. C. Runger, *Applied Statistics and Probability for Engineers*, John Wiley & Sons, New York, 2nd edition, 1999.

[3] J. S. Bendat and A. G. Piersol, *Random Data Analysis and Measurement Procedures*, John Wiley & Sons, New York, 1986.

[4] P. R. Bevington and D. K. Robinson, *Data Reduction and Error Analysis for the Physical Sciences*, McGraw-Hill, New York, 2nd edition, 1992.

Part IV

Engineering practice

Fig. IV.1. A construction site exemplifies the practice of civil, mechanical, electrical, and other engineering specialties. Project management is common to the practice of all engineering disciplines, and all engineers are responsible for ensuring public and worker safety.

Part IV. Engineering practice

Engineers, regardless of discipline, have something in common: the ability and responsibility to solve problems, particularly problems of engineering design. Part IV introduces

- the creative design process and how to stimulate creativity,
- the protection of intellectual property that results from design,
- the planning and scheduling of engineering projects,
- the assessment of the critically important need for safety in engineering, and
- the management of risk.

These key aspects of engineering practice are discussed in the following chapters:

Chapter 15—The engineering design process: Whether a design is totally different from a previous design or only a minor modification, the problem-solving technique followed by the design engineer usually follows a well-known series of general steps. This chapter describes the steps in a typical engineering design project and illustrates the process with a few simple examples.

Chapter 16—Creativity and decision-making: The synthesis step of engineering design described in Chapter 15 requires creative inspiration. This chapter describes methods of stimulating creativity and for making key design decisions.

Chapter 17—Intellectual property: Engineering work creates results that may qualify for legal protection as intellectual property. The principal types of intellectual property are patents, trademarks, copyright, industrial designs, and integrated circuit topographies. This chapter defines these types, explains how to protect them, and also how to use patent literature to stimulate ideas for future designs.

Chapter 18—Project planning and scheduling: This chapter briefly explains two basic, commonly used project-management tools: Gantt charts and the critical path method.

Chapter 19—Safety in engineering design: As an engineer, you will be responsible for the safety of any design that you approve. This chapter gives a basic introduction to safety requirements in engineering design, together with guidelines for eliminating workplace hazards. The importance of engineering codes and standards is also described.

Chapter 20—Safety, risk, and the engineer: This chapter further examines safety in engineering design and describes techniques for evaluating and managing risk in complex engineering projects.

Chapter 15

The engineering design process

Design is a fundamental activity that distinguishes engineering from disciplines based on pure science or mathematics. Engineers must have have solid, basic knowledge of these subjects but must also be able to apply their talents to create new products, processes, devices, and systems. For example, the computer mother board shown in Figure 15.1 is the result not only of a knowledge of physics, but of innovative design of computing circuits, components, chips, and devices on the chips. Good design needs more than just hard work; it requires creative thought. However, when you are faced with a design problem, where do you start? Design engineers normally follow a well-known series of steps to reach the goal. This chapter discusses

- a working definition of the word *design*,
- a descriptive outline of the design process,
- how to organize a design team.

Fig. 15.1. A computer circuit board illustrates design and optimization at the device, circuit, chip, and board level.

1 Definition and importance of engineering design

Engineering design may be defined in simple terms as the process of developing workable plans for the construction or manufacture of devices, processes, machinery or structures, to satisfy some observed need. A more formal definition has been published by the Canadian Engineering Accreditation Board [1]:

> "Engineering design integrates mathematics, basic sciences, engineering sciences and complementary studies in developing elements, systems and processes to meet specific needs. It is a creative, iterative and

often open-ended process subject to constraints which may be governed by standards or legislation to varying degrees depending upon the discipline. These constraints may relate to economic, health, safety, environmental, social or other pertinent factors."

Both of the above definitions emphasize that engineering design is the creative application of technical knowledge for some useful purpose. Engineering design is as creative as artistic design and differs only in the need for knowledge of science, mathematics, and technology to carry it out properly. Moreover, engineering design is vitally important. It is the source of the ideas, plans, and products that support healthy construction and manufacturing industries. In fact, the prosperity of an entire country is directly related to the value of goods designed and manufactured within the country and their sales relative to imported goods. A country blessed with competent designers can overcome a shortage of natural resources, as England showed after the Industrial Revolution and countries in Asia have shown since, whereas a country blessed with natural resources, such as Canada, will have difficulty competing internationally unless the creative spark of its engineers is encouraged and developed.

Design is therefore important to all of society and is the creative process at the core of the engineering profession. Real engineering is far more fun than reading about it, but you should take the time to learn the design process because it is a general problem-solving technique, and it will be useful, in class and in the workplace, during your professional career.

The following sections describe the steps of a design project. At each step, what to do is emphasized, rather than the details of how it should be done. This description is most applicable for individuals and small groups for which creativity and flexibility are of primary importance. In large organizations, where success depends less on individuals and clear communication of progress is crucial, a more detailed, formal model of the process is sometimes used, as described in textbooks such as [2].

2 The design process

Many people believe that design is a mysterious combination of luck and inspiration. Nothing could be further from the truth; good design is mainly a matter of organization, teamwork, and communication. In the design process, both creativity and criticism are essential but at different times, since uncontrolled criticism causes the creative spirit to dissipate. Figure 15.2 shows the steps in a typical design process. This process is also an effective problem-solving technique; it gives you an organized strategy to attack problems, even when they do not involve design. Therefore, you should review it until you understand it completely. A brief description of the seven typical steps in the design process follows.

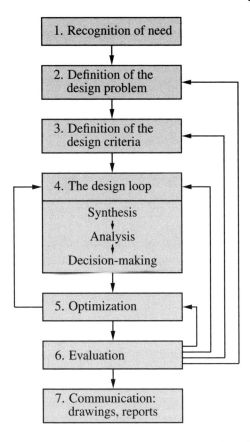

Fig. 15.2. The design process, showing the many possibilities for iteration.

2.1 Recognition of the need

Every design begins with a recognition that there is some need for improvement. Needs may be obvious, such as providing a water supply to a new housing subdivision, or they may be hidden and revealed only by investigations, surveys, or research. For example, research shows the need to solve the problems of global warming and ozone depletion. Needs are usually very vague; you may be told simply, "Do something about" a failed structure, a jammed production line, a pollutant, or an unprofitable product. However, design decisions cannot be made until the problem is defined more clearly.

2.2 Definition of the design problem

Defining the problem is sometimes a major task and usually requires gathering information about the problem and perhaps research or preliminary feasibility studies, in some cases, to ensure that the problem is defined in a way that is solvable. For example, the need to clean up a polluted lake may be obvious, but it would be foolish to begin designing a new sewage treatment plant if the problem is caused by industrial waste or excess lawn fertilizer.

In particularly complex designs, with many possible alternatives, an evaluation method, called the criterion function, is created, in which numerical values are assigned to each of the criteria for each alternative. The numerical values are then summed, and the design with the greatest sum is the optimum. This technique is discussed in the next chapter.

2.3 Defining the design criteria and constraints

While the problem is being defined, the design criteria and constraints must be identified.

- *Design criteria* are performance standards to be met by the design.
- *Design constraints* are limitations placed on the designer, the final design, or the manufacturing process.

For example, in the design of an automobile, the speed of the vehicle and its aerodynamic efficiency are design criteria; the budget limit for the design team is a design constraint.

Where possible, the criteria and constraints must be easily measured quantitative values, not subjective values that depend on opinions. For example, if the problem is to "design a new, quieter, motor exhaust muffler," then the criteria must include the maximum acceptable noise level in decibels and the distance and direction from which the noise will be measured; otherwise, it may be impossible to agree whether the final design is acceptable, since the designer's idea of "quiet" may not agree with that of the client.

Some design tasks do not easily allow the definition of purely objective criteria. Consider, for example, the problem of designing a "more convenient" wheelchair, a computer program that is "simple to use," or a web page that is "easy to understand." Measurement of success in these tasks requires the observation and quantification of human opinion and behaviour, which may introduce subjective factors. The introduction of computers throughout the workplace and home has increased the importance of this class of problems, which includes not only the design of workstations and web pages, but also all situations where automation affects and is directed by people, as in the control of aircraft, automobiles, home electronics, and large industrial plants. Human-interface design is a growing specialized discipline; representative references are [3] and [4].

Students on work-terms or internships are often asked to write computer programs for others to use. Do not underestimate the importance of defined criteria for measuring convenience of use, a factor that has often been ignored. For example, in a classic study of computers in business and industry [5], Landauer compared companies that had invested heavily in computers with those that had not, and found that the latter were generally more successful. He attributes this result to poor design of software used for service tasks. This situation may have changed since his study, but the fundamental conclusion, that user tests are crucial, still stands.

At the end of the problem definition stage, it must be possible to make a statement beginning: "Design a component (or system or process) to accomplish (a specified task), subject to the following design criteria (performance standards)... and subject to the following constraints (on the design, the design team, or the manufacturing process)..."

2.4 The design loop

Design is essentially a repetitive process of:

- *synthesis:* suggesting ideas or methods to solve the problem,
- *analysis:* calculating the expected result of each idea or method, and
- *decision-making:* deciding which alternative is best.

These three tasks are the key to good design and are discussed in more detail in the next chapter. As an illustration, consider the problem of designing a gas pipeline: During the synthesis phase, pipe diameters, pump sizes and pipe routes from sources to consumers would be suggested. In the analysis phase, the costs, flow rates, and construction problems associated with each combination would be determined. In the decision-making phase, the combination of sizes and routes that yields the best flow rate per dollar, with the minimum of problems, would usually be recommended for construction. If no combination were found to be acceptable, then the loop must begin again with better ideas, to achieve more creative synthesis, or the problem definition or criteria must be changed.

In virtually every design project, the cycle of synthesis, analysis, and decision-making is an iterative design loop, repeated many times. In his now-classic text, Asimow [6] points out that the design gets more complete and more detailed with each repetition. His experience is that there are at least three iterations of the design loop, which he calls feasibility study, preliminary design, and detail design. Each iteration may require an increasingly specific information search, creative synthesis, and analysis, perhaps including theoretical studies, computer simulation, and laboratory or other tests. However, these activities progress from very general project goals to very specific details, as follows:

- *The feasibility study:* The first iteration usually involves a quick synthesis of methods to solve the design problem and an analysis of these alternatives. The goal is to answer the question, "Is a solution possible?"
- *Preliminary design:* When at least one solution is considered possible, the preliminary design cycle can begin. The purpose now is to establish a general plan to guide the detail design. Again, this requires synthesis and analysis and may include project layouts or mock-up models. The goal is to answer the question, "What are all the main components of the design?"

- *Detail design:* In the third iteration, each detail of the design is subjected to synthesis and analysis. Drawings are prepared for fabrication, assembly, and construction, and every nut and bolt must be identified. This iteration usually ends with a prototype, or working model. The specifications for tenders, which are bids for manufacturing or construction contracts, are prepared, and decisions are confirmed for every aspect of the design. This iteration answers the question, "Have we forgotten anything?"

2.5 Optimization

The result of the process described previously is an acceptable design. However, the designers must ask themselves if it is the optimum design. The term *optimum* implies a compromise between costs and benefits; it is the best design that can be achieved at reasonable cost. To evaluate optimality, the proposed design is evaluated according to the predefined design criteria. Optimization improves the efficiency of the design; that is, a better design is obtained at lower cost.

2.6 Evaluation

When the design is completed, it is customary to hold a design review in which a senior engineer, and possibly the client, approves drawings and specifications before they are released for manufacturing or construction. The senior engineer must have the knowledge and experience to detect errors or omissions and the authority to approve or send drawings back to the designers for changes or further optimization. If an optimum design cannot be achieved, the senior engineer might help to revise the problem definition, the design criteria, or the constraints.

2.7 Communication of the design

When approved, the design is communicated by releasing the drawings and specifications to construction or manufacturing groups. One or two engineers will usually be assigned to deal with minor problems; however, the major decisions have been made and the drawings, reports, specifications, and calculations contain all the design information needed for manufacturing or construction. The design team is then assigned to another project.

Although no two design projects are ever precisely the same, these seven steps, or steps that are similar, are typically observed in every design process. Other authors have suggested slightly different design steps [6–9], but the differences are minor.

3 Two design examples

To illustrate the design process described above, consider the following illustrative examples. The first is simple and mundane; the second involves a large design team in a major company.

Example 1. A dripping tap.

You are awakened in the night by the persistent dripping of a tap in the bathroom. You attempt to go back to sleep, but the annoying sound continues and you decide to do something about it. In this situation you have to solve a problem, and the solution can be analysed as a process of design, as follows.

You have identified a need: to stop this noise so that you can go back to sleep. You enter the bathroom and find that a faucet will not shut off properly; you then define the design problem: to repair the offending faucet by replacing its rubber washer. The design criteria are clear. The project must be done quickly, in less than 20 or 30 minutes, say, and effectively: the faucet must not leak. However, there are constraints on your solution. You must use the tools at hand, and you must not waken the landlord because you are overdue on your rent.

On your first attempt at synthesis, you visualize three possibilities: call a plumber, ask the landlord to do it, or repair it yourself. After estimating the cost and effectiveness for each possibility, your analysis indicates that the plumber would be effective but would take time and money; the landlord would be awkward and probably ineffective, but if you repair it, the job will be cheap and fast. Therefore your decision is to repair it yourself.

Now you must repeat the design loop at a more detailed level. To carry out the job, you need wrenches and a new washer. You have these in your toolbox, so the job is feasible; you are now ready to go through the loop once again in more detail. You imagine (synthesize) a scheme to turn off the water, use the wrench to dismantle the faucet, take off the old washer and put on the new washer, replace the faucet, and turn on the water again. However, when you analyse this proposal, you realize that the main shutoff valve for the water is in the basement and you cannot reach it without waking the landlord. You decide that this proposal will not work.

At this point, you reach a dead end, or, as engineers used to say, "Back to the drawing board!" You must go back to the definition of the problem, and decide that instead of repairing the faucet, the problem could be redefined to construct a sound-proof barrier to eliminate the noise. You now generate ideas to solve this problem and arrive at a new proposal: Put cotton in your ears, which is uncomfortable, and close the bathroom door. You now review the design criteria and realize that all of the criteria would be satisfied at minimum cost and minimum time and, although you are slightly uncomfortable, you will be able to sleep. Therefore, this is the optimum design and you decide to carry it out.

The above example, trivial as it may seem, illustrates the design process and also emphasizes the repetitive nature of design and the fact that problems encountered at one stage may affect decisions made earlier, rendering them useless. It also shows that an optimum design may not be a perfect design, but it is the best compromise.

Example 2. Automobile design.

Consider a more complex example of the design process: an automobile manufacturing company wants to compete more effectively in the small-car market to increase profits (see Figure 15.3). A simplified description of the design process follows.

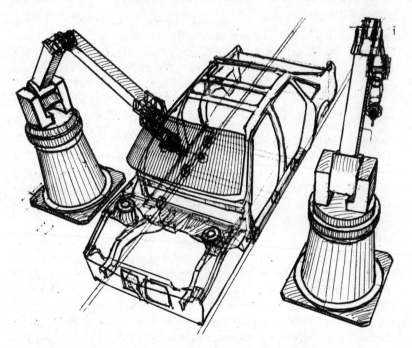

Fig. 15.3. Sketch of an automobile assembly station.

The need is already recognized; however it is very vague. Assume that after some study, the problem is defined more specifically "to design a new, efficient, low-cost small car." At this point, design criteria must be very formally stated. The design criteria in a large project are usually bound into a manual for the use of the engineers. For business reasons, such manuals are strictly confidential.

The criteria include power, performance, approximate size and weight, number of passengers, and other characteristics expected in the automobile. These criteria depend on the anticipated market. The design is also subject to constraints such as government standards for highway safety and air pollution controls. The constraints may also include preferred dimensions and production line data, so that stocks of standard parts, jigs, fixtures, and robots can be used in manufacturing.

The key creative process, or synthesis, then begins. Methods must be suggested for increasing efficiency and reducing cost. These may include aluminum engines for lower weight, front-wheel drive for less drive-train friction, fibreglass bodies for lower weight and cost, and many other possibilities. The feasibility of these possibilities must be checked by gathering information and conducting tests; that is, by analysis. For example, the aluminum engine concept would require studies of the thermodynamic engine cycle and heat dissipation, metallurgical studies of the expected wear and corrosion in the cylinders, lubrication studies of bushings, flow studies of the cooling system, and other items. A deadline must be set for deciding if the initial concept is feasible.

Assuming that the decision to continue is made, preliminary design work begins. Layout drawings of the automobile are made, and improvements are suggested to the best ideas from the feasibility study. Analysis calculations are carried out to check engine power, performance, stability, ease of manufacturing, and other factors. "Mock-up" models of the auto are usually built to check styling, seating, clearance between parts, and ease of repair. The synthesis and analysis is repetitive: analysis leads to suggestions for better synthesis of ideas, which are analysed in turn. The cycle continues until an optimum design is obtained.

The final detail design begins when conflicts have been eliminated from the preliminary plans. Every part in the automobile must be defined, analysed, and specified. The procedure for manufacture and assembly must be decided. As the detail design nears completion, the manufacturing personnel on the design team release advance drawings so that new equipment can be designed and built for the production line.

When the detail design is complete, drawings and specifications are given a final evaluation, or design review, before being released for fabrication and assembly to begin. The design team has now completed its task. After a short vacation, and probably before the first small car reaches the end of the production line, design will begin for the following model.

4 Design team organization

Most engineers work together on projects as members of design teams. Design teams are similar to sports teams: there are top scorers and bench-warmers, and it takes a good coach and team spirit to achieve a winning season. Some basic rules that the senior engineer, who may function as "coach," should follow, in organizing a design team for a specific project, are listed below:

- Avoid groups greater than seven or eight persons; they are too large for easy communication.
- Arrange the work so that each person has a portion of the project over which he or she exerts authority and bears some responsibility.
- Ensure that each team member understands how his or her part of the design contributes to the main objective.

- Ensure that schedules, deadlines, and intermediate objectives are well understood. Monitor deadlines with encouragement, not harassment.
- Encourage creativity.

It has been observed that there are seven critical functions that are carried out by members of a successful design team. The tasks in the design process can be categorized according to these functions. The functions, listed below, give an alternative insight into the design process and the personalities required on a successful design team:

- *Generating ideas:* synthesis
- *Supporting ideas*
- *Encouraging people*
- *Gate-keeping:* keeping abreast of information related to the project and communicating it to team members
- *Problem-solving:* The key steps that require problem-solving ability are defining the problem, defining criteria, and analysis.
- *Controlling the quality of technical decisions:* a combination of decision-making and optimization
- *Managing the project:* personnel selection, planning, and scheduling

Very few people have the ability or personality to perform all of the above critical functions. Therefore, it is important to encourage diversity in selecting members of a design team. For example, a group composed entirely of excellent problem solvers might have difficulty generating good ideas to analyse. All of the above tasks are equally important, although they are not equally visible. For example, the team member who suggests an essential idea or solves a difficult problem may have received information or encouragement from another team member. All roles need recognition, and a good coach rewards those who assist as well as those who produce major results.

5 Further study

1. Assume that you are the Personnel Officer of a large engineering company and you want to hire some very creative junior engineers to work in engineering design. What simple questions would you ask a prospective employee to get an indication of creative ability? What answers would you expect from the best applicants?

2. The third step in the typical design process is to define the criteria to be satisfied by the final product and the constraints on the project, the design team, or the manufacturing process. In large projects, these criteria would be bound into a manual for easy reference by the engineering team. For each major project listed below, suggest at least 10 topics or headings that would probably be included in the design criteria/constraint document.

(a) a passenger aircraft,

(b) a military personnel carrier,

(c) a tractor-trailer truck,

(d) a personal computer,

(e) a city water distribution system,

(f) a sewage disposal plant,

(g) an electric power distribution grid,

(h) an electric transformer station,

(i) a portable furnace to burn PCB (toxic) chemicals.

3. The design process usually begins with the recognition of a need. As explained in this chapter, the need may be vague or poorly expressed, and it is the responsibility of the engineer to define the problem clearly and follow the design process until workable plans are developed for satisfying the need. For the needs listed below, carry out the design process for each project, to the extent specified by your professor. Devise a new or improved item from the selection below.

(a) jack for lifting automobiles,

(b) brake for bicycles,

(c) system for airport luggage handling,

(d) silent alarm clock,

(e) device for recycling and purifying household water,

(f) winch for hauling logs (for cottages),

(g) speed/gear changer for bicycles,

(h) desk for engineering students,

(i) mechanism for locking windows,

(j) shock-absorber system for auto bumpers,

(k) electrical wall outlet,

(l) street layout or traffic pattern for your campus.

6 References

[1] Canadian Engineering Accreditation Board, *Accreditation Criteria and Procedures for the Year Ending June 30, 2001*, Canadian Council of Professional Engineers, Ottawa, 2001, <http://www.ccpe.ca/files/report_ceab.pdf> (March 25, 2002).

[2] N. Cross, *Engineering Design Methods*, John Wiley & Sons, Inc., New York, 3rd edition, 2000.

[3] C. D. Wickens, S. E. Gordon, and Y. Liu, *An Introduction to Human Factors Engineering*, Addison-Wesley, New York, 1998.

[4] D. Wixon and J. Ramey, Eds., *Field Methods Casebook for Software Design*, John Wiley & Sons, New York, 1996.

[5] T. K. Landauer, *The Trouble with Computers*, The MIT Press, Cambridge, MA, 1995.

[6] M. Asimow, *Introduction to Design*, Prentice Hall, New York, 1962.

[7] J. D. Kemper, *Introduction to the Engineering Profession*, Saunders College Publishing, New York, 2nd edition, 1993.

[8] S. Love, *Planning and Creating Successful Engineered Designs*, Advanced Professional Development, Inc., Los Angeles, 1986.

[9] J. W. Walton, *Engineering Design: From Art to Practice*, West Publishing Company, New York, 1991.

Chapter 16

Creativity and decision-making

The design process described in Chapter 15 is a useful outline of a typical design project. The most difficult part of this process is probably the iterative loop of synthesis, analysis, and decision-making. Design engineers occasionally run into a mental block during the synthesis step, and a better knowledge of the creative process may help overcome this barrier. Also, even when many feasible designs are suggested, the designer may have difficulty choosing the best design. This chapter describes methods of coping with these problems by

- dissecting the creative process into its component parts,
- listing several methods for stimulating creativity, and
- describing two simple methods for making difficult design decisions.

1 Synthesis: The creative process

Synthesis is simply another name for creative inspiration. It is the process of putting together the ideas in your head to create a solution for the problem that you are trying to solve. Synthesis is obviously very important in engineering. However, sometimes synthesis is difficult, tricky, or unpredictable. To help you better understand the creative process of synthesis, it might help to observe that it can usually be broken down into five separate stages:

- *Gather information:* Before any solutions can be generated, full information must be gathered about the problem. What is the "real" problem? Have we defined it properly? What is the history of the problem? How are similar problems solved? Do we have all of the knowledge or skills needed to solve the problem?

- *Make a concentrated mental effort:* Using the information that you have gathered and your knowledge of similar problems in the past, try to generate a solution to the problem. If this fails, gather more information, seek advice from others, and try again. Experiment with techniques for stimulating creativity, such as brainstorming, discussed later in this chapter. Immerse yourself in the problem, and work hard to find a solution. If no solution is satisfactory, then make a list of the best "partial" solutions that you have.

- *Give it a rest:* If you still can't solve the problem, after you have made a serious effort to do so, then relax, put the problem away, and "sleep on it."

Your subconscious will continue to work on the problem, even though you may not be aware of this mental activity.

- *Revelation:* In most cases, when you return to the problem, you will find that the solution is now apparent. While you were resting, whether the rest was an hour, a day, or a week, your subconscious sorted out the pieces to the puzzle, and the solution became clear.
- *Implementation:* When you finally receive the revelation or insight, check that it is sufficiently thorough and general, then put it into action.

In a few cases, the revelation never comes, and you must decide whether to try again or to abandon the effort. Examine the problem to see if it is properly stated. If it still seems well-posed, then start again at the first step by gathering more information.

Many great inventors state that solutions came to them unexpectedly in a moment of leisure. For example, Newton is alleged to have discovered the law of gravity when he was hit on the head by a falling apple while sitting under an apple tree. However, such anecdotes never mention the years of preparation and concentrated mental effort that preceded the revelation.

Synthesis is not a single instant of inspiration; it is a process. If you understand the process, and if you are willing to put the effort into gathering the information, making a concentrated effort, then relaxing to let your subconscious mind work, then you may be surprised to find, perhaps when you least expect it, when sitting in your bath or eating an apple, that a revelation will come to you and your problem will be solved.

2 Characteristics of creative people

Everyone has a creative instinct; however, it is stronger in some people than others. What is the key characteristic that makes a person creative? Obviously, a creative person needs a certain amount of education and intelligence to understand ideas and to put them together. However, many innovative people have little schooling and would rank low in standard intelligence tests, so these are not the key characteristics. Similarly, experience and age would seem to be good characteristics because experienced, older people have faced and solved problems previously. However, we have all met, at one time or another, an experienced older person who is "in a rut" with no innovative or creative urge.

The same arguments apply to personality characteristics. People who are well organized seem to get more done in an average day, but is the work really creative? Many of our most creative friends have lives that appear to be in total disorder, but disorder, itself, surely is not a key characteristic in creativity.

In his excellent inspirational book on creativity, Roger von Oech [1] cited a psychological study carried out to identify the key characteristic of creative people. The study, which was commissioned by a major oil company and conducted by qualified psychologists, came to the simple conclusion that creative people thought they were creative, and less-creative people didn't think they were creative. In other words, people who think they are creative and want to be creative subconsciously encourage creative ideas, whereas people who don't think they are creative or, for whatever reason don't want to be creative, stifle their creative ideas. The key to creativity is the desire and the conviction that you are creative.

Therefore, if you are a student of engineering, then you have the education and intelligence to be a creative person. Experience is important, of course, but you will get that as the years pass. The most important characteristic for you to encourage is the goal of being creative. If you want to be creative, and if you think you are creative, then you will be creative.

3 Stimulating engineering creativity

The creative process described above usually leads to a successful solution, but when it fails, there are ways to stimulate creativity and to overcome mental barriers. The first way to stimulate creativity is simply to discuss the problem with other engineers. If no suitable colleagues are available, it is sometimes productive to explain the problem to a non-engineer. The act of explaining your problem requires you to state it clearly, and this simplified statement may trigger a solution in your mind. If you simply have no colleagues available, you can write the problem on a sheet of paper, along with a list of the best "partial" solutions that you have, so far. This act may stimulate a solution, but even if it doesn't, you now have the problem clearly defined, and perhaps it is time to take a rest, as suggested above, and let the idea incubate in your mind for a while.

An obvious way to get more solutions to a problem is to put more people to work on it. However, unless they are well-managed, group efforts at generating ideas do not usually increase the speed and efficiency of the creative process. The reason for this curious anomaly is that personal inhibitions and personality conflicts prevent the free flow of ideas. To be effective, group efforts at stimulating ideas must encourage team spirit while controlling the negative criticism of colleagues. The group techniques described below are the two most common methods of many suggested in the references [1–4].

3.1 Brainstorming

A group of people, usually four to six persons, is presented with a problem statement and asked to consider the facts presented. The solution process then follows two distinct phases:

- *Synthesis:* In the first stage of creative synthesis, group members are asked to suggest possible solutions. No critical, negative, or judgemental comments are permitted. The group leader must enforce this "no-criticism" rule strictly. All suggestions, no matter how weird or unusual, are written down.

At the end of a fixed period of time, usually 20 to 30 minutes, the synthesis phase ends and analysis begins.

- *Analysis:* The analysis involves criticizing and evaluating the list of ideas and ranking them in order from most practical to least practical. When the ranking is finished, the group then examines the most promising suggestions in more detail.

Brainstorming is effective and is widely used. It has several variations, and for difficult problems, it may be more productive to ask each person to give the problem some deeper thought over the next hour, day, or week, then disband the group and re-convene later.

3.2 Brainwriting

The brainwriting process is almost identical to brainstorming, except that it involves written rather than verbal communication. The group leader must prepare tentative solutions or suggestions prior to the meeting on separate sheets of paper. These sheets are placed in the middle of the table at the start of the meeting. After the facts of the problem have been stated, each person selects a sheet from the centre of the table, reads it, and tries to improve it or to suggest alternative solutions. When you run out of ideas, you exchange sheets with one another and continue to try to improve on the ideas suggested. All participants exchange constructive ideas this way, but negative comments are effectively suppressed. The notes serve as a permanent record of the meeting. Analysis or judgement of the ideas takes place later, when the ideas are ranked and examined in detail, as for brainstorming. Brainwriting is suitable for email, and many internet chat rooms are, in fact, a form of brainwriting.

3.3 Overcoming creative blocks

In his brief but entertaining book, von Oech [1] describes 10 blocks to creativity and provides dozens of exercises to overcome them. According to von Oech, when it comes to creativity, many people mistakenly believe one or more of the following:

- There is only one right answer.
- The creative process must be logical.
- They must "follow the rules" even if the rules are unwritten.
- They must be practical and therefore inhibit their fantasies.
- They must avoid ambiguity and therefore stifle their imagination.
- They avoid new ideas for fear of making mistakes.
- Play is frivolous, and new ideas are hard work.
- They narrow their focus and miss ideas in nearby areas.

- They are afraid to look foolish by suggesting an unworkable idea.
- They are not creative.

Of course, all of these ideas are wrong. There are many ways to solve most problems, and the creative process involves ambiguity, imagination, playfulness, and enjoyment. It is important to have a positive attitude. We must not be afraid of looking foolish, and we must avoid applying unwritten rules that inhibit our playful creativity.

The above suggestions for stimulating creative thinking may appear unusual. However, when problems occur in engineering projects and time is valuable, these methods sometimes trigger a breakthrough.

4 Analysis

The dictionary defines analysis as the separating of a whole into its component parts, which is the opposite of synthesis. In engineering design, analysis means the application of mathematics or logic to determine whether the design will function as required. Analysis is sometimes called *deductive* or *convergent thinking*, since it begins with the application of a potentially infinite set of principles and arrives at a single conclusion. Synthesis, on the other hand, is called *divergent thinking*, since it starts with a single problem and ends with a potentially infinite set of solutions.

Analysis, and the mathematical and logical thought processes that accompany it, forms most of the course content of most engineering curricula; there is a lamentably small opportunity for creative thinking until students enter the real world of engineering, where creativity and initiative are essential. Consequently, there is no need to discuss analysis further; engineering students are already well acquainted with it.

5 Decision-making

In engineering design, decision-making always follows the synthesis and analysis steps. For example, consider the design of a heating system for a building. Assume that two suggestions have been made: electrical heat and natural gas. These alternatives may be analysed to get costs of ductwork versus wiring, total energy costs, and expected maintenance costs. The alternative with the lowest cost is chosen. However, what happens if we wish to include other factors, such as environmental considerations, future variations in energy supply, and future building enlargement? In some cases, electrical heat may be superior; in other cases, natural gas may be superior. How do we choose? In other words, how do we arrive, logically, at the best decision, when many factors are involved but are not equally important?

It is possible to discuss only very basic decision-making concepts here; further reading is available in the references. The simplest case is decision-making under certainty; that is, when the result, or outcome, of each decision can be determined in advance. In the example given above, all of the outcomes must be known, including the future variation in energy supply. If any factors are not known, then estimates must be made.

5.1 A basic decision-making method

When faced with an important decision, engineers typically do the following:

1. List all of the possible choices or courses of action.
2. List, separately, all of the possible factors or criteria that could affect the decision.
3. Compare the two lists. If any choices are obviously not practical, remove them from the list.
4. List the advantages and disadvantages for each choice. All of the criteria should appear in the lists. The best decision is usually obvious at this point as the choice with the most advantages and the fewest disadvantages.

If the steps above have been followed and no decision is obvious, then the computational decision-making method, below, may be useful.

5.2 Computational decision-making

Decision-making, when done systematically, requires a thorough comparative evaluation of all possible decision alternatives, which are assumed to be given or to have been defined by using one or more of the creative methods described earlier in this chapter. The evaluation of alternatives is aided by quantitative techniques such as the one described below.

A numerical payoff value is calculated for each alternative, and the alternative with the largest payoff is chosen. Alternatively, a cost function to be minimized can be calculated, but these two methods are equivalent in this context since a positive cost is a negative payoff. Payoff will be used in the following. The general method will be described, followed by an example.

Suppose there are m alternatives. Then the payoff values denoted f_i, $i = 1, 2, \cdots m$ must be calculated.

The payoff for each alternative will be evaluated according to a set of n criteria, and each criterion will be assigned a relative importance, or weight, w_j, for $j = 1, \cdots n$, such that the weights sum to 1, or to 100 %. For each alternative, the total payoff will be defined to be a weighted sum of n partial payoffs associated with the criteria.

For each criterion, the alternatives will be assigned coefficients that estimate how well they satisfy the criterion. The maximum coefficient is exactly 1. Thus, for each criterion j, we assign a fraction p_{ij} for each of the alternatives $i = 1, \cdots m$. Then the total payoff for alternative i will be the weighted sum

$$f_i = \sum_{j=1}^{n} p_{ij}\, w_j. \tag{16.1}$$

A variation of the above method that may be simpler to interpret is to use an arbitrary numerical scale for the coefficients associated with criterion j and then to divide by the largest of their absolute values. Thus, for each j, assign coefficients c_{ij}, $i = 1, \cdots m$, find the maximum absolute coefficient

$$C_j = \max\{|c_{1j}|, |c_{2j}|, \cdots |c_{mj}|\}, \tag{16.2}$$

so that $p_{ij} = c_{ij}/C_j$, and f_i is calculated as

$$f_i = \sum_{j=1}^{n} \frac{c_{ij}}{C_j}\, w_j. \tag{16.3}$$

Then Equation (16.3) and Equation (16.1) produce identical payoff values f_i.

The difficulties associated with a quantitative method are in determining reasonable weights and precise coefficients. However, even if confidence in the numerical values is lacking, this formal method requires all of the relevant factors to be clearly stated and carefully evaluated.

The computations can be performed systematically by using a chart or table, but they are suited naturally to computer spreadsheet software, as illustrated by the following simple example.

Example 1. Student travel.

Consider the case of a student living in a rented room that is about 2 km from the university. It is necessary to travel this distance twice each day, and the student must decide which mode of travel is best. There is no bus service, so the choices are walking and riding a bicycle. However, it is conceivable that the student might buy a motorcycle or automobile if it is not too expensive. The student applies the formal decision-making process in order to compare the modes of travel. The criteria are operating cost, time, and safety, which are weighted 30 %, 40 %, 30 %, respectively; thus time is slightly more important than the other two criteria. Table 16.1 summarizes the computations.

The entries in the table are computed as follows:

- Operating cost is in the first column: walking has zero cost. The car is considered to be twice as expensive as the motorcycle and 10 times as expensive as the bicycle. The values are negative since costs are negative payoffs. The computed maximum absolute value is 10.

Table 16.1. Payoff computation for the student-travel problem. The four alternatives have been compared using three criteria: operating cost, travel time, and safety. The bicycle has the most positive payoff or, alternatively, the least cost.

	Coefficients			Payoff			
	Cost	Time	Safety	Cost	Time	Safety	Total
Walking	0	−35	0	0	−40.0	0	−40.0
Bicycle	−1	−20	−1	−3	−22.9	−6	−31.9
Motorcycle	−5	−8	−5	−15	−9.1	−30	−54.1
Car	−10	−8	−2	−30	−9.1	−12	−51.1
Max abs val	10	35	5	30	40	30	100

- Time is in the second column. Walking takes 35 minutes, the bicycle takes 20 minutes, and the car and motorcycle each take 8 minutes. All are entered negative, and the largest absolute value is 35.
- A somewhat arbitrary safety estimate is in the third column. Walking is neutral, but the motorcycle is judged to be five times as dangerous as the bicycle, with the car somewhere in between.
- The partial payoffs are computed in the next three columns. For example, each partial payoff in the Cost column is the corresponding Cost coefficient, divided by 10, the largest absolute column value, and multiplied by 30%, the Cost weighting factor. The partial payoffs are summed in the rightmost column.

Table 16.1 shows that the alternative with the least cost is the bicycle. Therefore, based on the choices available, the preferences indicated by the weightings, and the coefficients assigned, the bicycle would appear to be the best mode of travel. This decision may change if any of the numerical values change; for example, the safety of the bicycle in winter weather may not be as favourable.

6 References

[1] R. von Oech, *A Whack on the Side of the Head; How to Unlock Your Mind for Innovation*, Warner Books, New York, 1983.

[2] H. S. Fogler and A. E. LeBlanc, *Strategies for Creative Problem-Solving*, Prentice-Hall, Upper Saddle River, NJ, 1995.

[3] V. J. Papanek, *Design for the Real World: Human Ecology and Social Change*, Thames and Hudson, London, 1985.

[4] J. W. Walton, *Engineering Design: From Art to Practice*, West Publishing Company, New York, 1991.

Chapter 17

Intellectual property

Engineers create designs, processes, ideas, and products, and when this creativity is expressed in tangible form, it becomes intellectual property.

In a broad context, the term *intellectual property* refers to the ideas and creations produced by others and ourselves. How we use and acknowledge the ideas of others is an ethical matter, as discussed in Chapter 3, and how to acknowledge their published written work is described in Section 1.2 of Chapter 8. However, specific laws, as well as ethics, apply to intellectual property that has been put into tangible form. We are concerned with the legal definitions of intellectual property, our rights to protect what we own (see Figure 17.1), and how to obtain and use the intellectual property of others.

In this chapter, you will learn

- the importance and use of the formally defined types of intellectual property,
- the types of intellectual property for which special laws have been written,
- how to obtain protection for your intellectual property.

Fig. 17.1. Engineers have a sense of humour. These microscopic designs, and many others [1], have been discovered on commercial microcircuits. Initially such patterns provided protection against illegal chip copying (see Section 7), but now they are an expression of the chip designer's individuality. Some companies allow only the designer's initials to be used. (Courtesy of Chipworks, Inc., Ottawa, Canada)

1 Introduction

The intellectual property considered in this chapter is, generally, any creative material that has been put into tangible form. Pure ideas do not qualify, but when they are recorded or written, for example, the recordings or written copies are intellectual properties.

The following sections first contrast proprietary property, which is owned by an individual or company, with public-domain property, which is accessible by all. Then the principal types of intellectual property will be described, with a brief description of how the rights of their owner are established. The remainder of the chapter gives more detail about each of the types of intellectual property.

1.1 Proprietary intellectual property

Information for which ownership has been established is said to be *proprietary*. The owner may be an individual or a legal entity, such as a corporation or partnership. Intellectual property can be owned, bought, sold, or shared, just like any other tangible asset.

Government legislation sets the legal rules regarding intellectual property. As a result, legal protections such as patent law do not extend outside the country that passed the law. However, governments sign agreements and conventions to extend proprietary rights between countries.

The following six types of intellectual property are of interest to engineers; Table 17.1 summarizes their properties and the laws that govern them:

- patents,
- copyrights,
- trademarks,
- industrial designs,
- integrated circuit topographies,
- trade secrets.

The formal process of establishing ownership is similar to the process for land; an application is made to a government office, which registers ownership of the property if the applicable legal requirements are met. Copyright protection, however, does not require formal registration. Once registered, a description of the property and the registration documents are public. The Canadian Intellectual Property Office (CIPO), which is part of Industry Canada, registers the first five types of property in the above list. Trade secrets are not registered and do not require disclosure or have any special protection.

The types of intellectual property listed above are defined and described in the sections that follow. However, all definitions and interpretations are subject to the re-interpretation of lawyers and courts and are changing rapidly in some areas, as pertinent laws and practices prove to be out-of-date.

Two examples of rapid change will be mentioned. First, electronic media make it very easy to freely distribute many kinds of artistic works, and the applicability of copyright to electronic documents is under active debate. Second, existing laws are difficult to apply to recent developments in bioengineering. These

Table 17.1. Type, criteria for protection, and term of intellectual property.

Intellectual Property	1. Type of legal protection 2. Criteria for protection 3. Term of protection
Patent	1. The right to exclude others from making, using, or selling the invention. 2. The invention must be new, useful, and ingenious. 3. 20 years from date of filing an application.
Copyright	1. The right to copy, produce, reproduce, perform, publish, adapt, communicate, and otherwise use literary, dramatic, musical, or artistic work, including computer programs. 2. The material must be original. 3. The life of the creator plus 50 years to end of the year; except for photographs, films, and recordings (50 years total).
Industrial design	1. The right to prevent competitors from imitating the shape, pattern, or ornamentation applied to a useful, mass-produced article. 2. The shape, pattern, or ornamentation must be original, and the article must serve a useful purpose. 3. Up to 10 years from registration.
Trademark	1. Ownership of a word, symbol, or design used to identify the wares or services of a person or company in the marketplace. 2. The trademark must be used in business, in Canada, before it can be registered and must not include prohibited words: profanity, geographic names, ... 3. 15-year periods, renewable indefinitely.
Integrated circuit topography	1. Ownership of the three-dimensional configuration of layers of semi-conductors, metals, insulators, and other materials on a substrate. 2. The configuration must be original. 3. 10 years from application.
Trade secret	1. Possession of a secret process or product. 2. Employees must sign non-disclosure contracts, agreeing to maintain the secret. 3. Potentially unlimited. However, disclosure or theft must be prosecuted under tort or criminal law.

and other cases require both new legal definitions and a re-examination of ethical conduct.

1.2 The public domain

Information that has no ownership is said to be in the *public domain*, which means that anyone can use it. Property in this designation includes material

- that is common knowledge,
- for which legal protection has expired,
- that has been placed in the public domain by the owner.

Most of our knowledge is in the public domain, including mathematical equations, natural laws, and information published in engineering, professional, and scientific journals. Newton's second law, for example,

$$f = ma, \tag{17.1}$$

where f, m, and a are force, mass, and acceleration, respectively, is public-domain knowledge.

Standard time is an example of published engineering information becoming public domain. Sandford Fleming was a railway engineer who helped to build the trans-Canada railway at the time of Confederation, when every station was on its own local sun time. When he missed his train, he decided to eliminate the confusion. Within three months he had published his first paper on the concept of standard time zones and the 24-hour clock. The eventual result was the time zone system we use worldwide today, in which, by the calculation $360°/(24\,h) = 15°/h$, every 15 degrees of longitude corresponds to a one-hour increment in standard time. Once he published this idea, it was in the public domain and, therefore, everyone's property. The system was adopted worldwide in 1884, with the primary meridian (0° longitude) passing through Greenwich, England.

A publication—a history book, for example—may contain common knowledge, but a particular written version of it may be the subject of copyright.

When legal protection of intellectual property expires, the property is in the public domain. Shakespeare's plays, for example, are in the public domain.

2 The importance of intellectual property

There are two important consequences of the principle of ownership and of the requirements for registering intellectual property:

- the owner can control and possibly profit from the use of the work by others;
- the records of registered intellectual property are vast public storehouses of knowledge.

The first purpose of protecting intellectual property is to ensure that its producer is rewarded for creating it. This protection encourages creativity, by the argument that few people would invest in research, development, or creation if they could not have exclusive use of the results. Under Canadian law, the owner of an intellectual property is permitted to benefit from it and to prevent competitors from using it. The owner may sell the intellectual property or may license it to others to manufacture or use.

A company's intellectual property may be its most valuable asset. For example, the rights for an integrated circuit chip might be worth more than the factory in which it is being manufactured.

It is up to the owner of the intellectual property to initiate proceedings to protect registered property; the Canadian Intellectual Property Office (CIPO) does not protect owners except by registering ownership.

As a practising engineer, you must know how to protect intellectual property; you must also understand and respect the property rights of others. Practising engineers also derive great value from the CIPO as a source of information about current technological developments. Advice about intellectual property can be obtained from:

>Publications Centre
>Canadian Intellectual Property Office
>Industry Canada
>Place du Portage I
>50 Victoria Street
>Hull, Québec, K1A 0C9

Most of the CIPO files and databases are open to the public [2]; this information permits you to monitor the progress of competitors and to inspect their intellectual property, as an aid to inspire your inventiveness. Even if you should find that your proposed patent, copyright, industrial design, trademark, or circuit topography is already owned by others, you may be able to license it from the owner, or, when the protection expires, you may copy and use it without permission.

2.1 Rights of employers and employees

The relationship of an engineer to an employer must be considered. Who owns the rights to intellectual property developed as part of employment activities? What about ideas developed using company facilities in the employee's spare time, unrelated to regular tasks? When an employee changes employers, how much knowledge can be taken to the new employer? The answers to these questions depend on individual circumstances, particularly the employment contract.

Employment contracts sometimes state that the company owns *any* intellectual property developed by the employee, who cannot publish or exploit intellectual property without the consent of the employer. Other employment contracts limit the company's claim to property related to the job. In many cases, companies are prepared to assume the effort and expense of registering the property for the mutual benefit of company and author. Hiring people away from a competitor is sometimes regarded as a form of industrial espionage (see Problems 4 and 5). An employment contract may prohibit an employee from working for a competitor of the employer for a period, two years, for example, after leaving the company.

The relationship of publisher to author is similar to that of employer to employee. Technical societies, which publish large amounts of engineering work, have created policies for ownership; see [3], for example.

3 Patents

A patent is the legal right, lasting 20 years, to exclude others from making, using, or selling an invention. For that time period, a patent is a piece of legal property. The first patent in Canada was granted in 1824, for a washing machine. The one-millionth patent was granted in 1976 for biodegradable plastic, useful for garbage bags. The Canadian Intellectual Property Office handles approximately 35 000 patent applications a year.

After the protection of private property, the second major function of a patent is *public notice*. An application for a patent requires a full and clear description, so that a skilled person could reproduce the invention [4]. In return for full disclosure, the owner receives the exclusive right to benefit from the use or sale of the invention. The patent records are a vital resource to industrial and academic institutions. However, the Patent Office does not enforce the patent law. This means that the protection offered by the patent can only be invoked by the patent holder through legal processes, which can be very expensive and time-consuming.

The principal patent legislation is the Patent Act of 1989. Earlier patents are regulated by slightly different rules. The term, or length of protection, is 20 years for a patent application filed after October 1, 1989. The Patent Act provides for the rights to an invention to be sold or licensed to others by registering a written document with the Patent Office.

A patent application may be submitted for any of the following, provided it is new, useful, and ingenious:

- a product (of manufacture),
- an apparatus (a machine for making other objects),
- a manufacturing process,
- a composition of matter,
- an improvement for any of the above.

In fact, the last category contains the largest number of applications.

The three key criteria—novelty, usefulness, and ingenuity—are examined for each patent application. The invention must be original, have a useful purpose, and must not be an obvious improvement over prior known art. These criteria require judgement. Patents granted in the past sometimes appear naïve or bizarre; see Figure 17.2, for example.

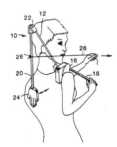

Fig. 17.2. Described as a self-congratulatory apparatus, this back-patting device received U.S. patent number 4,608,967 in 1986.

3.1 Patent developments

The first known patent to be awarded was to Filippo Brunelleschi by the Republic of Florence in 1421, while the first patent law protection for inventors was in the Republic of Venice, 1474 [5]. These events during the Italian Renaissance are evidence of the emerging concept of intellectual property and of the application of the new sciences to applied problems.

Perpetual motion machines have long been the goal of amateur inventors. As recently as 1979, a patent application was made for an "energy machine" that produced more energy than it consumed. Such inventions clearly violate the law of conservation of energy. More subtle attempts conserve energy but violate the second law of thermodynamics, proposing to rectify noise voltages or to drive a racket from molecular motion (see [6], for example, for a discussion of "Maxwell's demon.")

An invention may not be patented if it is intended for an illegal purpose or if it is already in the public domain. Pure ideas, scientific principles, and abstract theorems are not patentable. A process with a computer program embedded in it may be patentable, but computer programs themselves, even as algorithmic processes, are not patentable, although they can be protected by copyright as discussed in Section 4.

A patent application may fail if it depends on *prior art,* which means that it represents ideas and knowledge already in the public domain.

Genetically engineered life forms are patentable, thus protecting the investment made in the development of new biological organisms. For example (see [5]), a microbe for oil cleanup was patented in 1980, and a patent for a genetically engineered mouse appeared in 1988. Biotechnology is a new, complex, and disputable patent area; for example, a four-year litigation involving a drug for multiple sclerosis included a claim of over one billion dollars.

Biotechnology and software are two evolving areas where patent criteria are anything but clear. Moreover, the long patent process and the clarifications produced by legal challenges are sometimes outpaced by these rapidly changing technologies.

3.2 The patent application process

In principle, a patent application can be prepared and submitted by the inventor, and guidelines are available from the Patent Office [4]; however, preparing the application is complex, and patent agents are usually retained for this purpose.

The four main steps in the patent process are:

- retain an agent, and search existing patents,
- file an application,
- request an examination,
- protect the patent.

The first step in the patent process is the search of existing records to see what is already registered or known. Because of the complexity of the search, a patent agent or attorney is usually retained; their fees can amount to several thousands of dollars. The agent is typically not an expert in the subject matter of the patent and must be given careful descriptions and statements as a basis for preparing the application.

The patent application must be submitted within one year from the date of the first public disclosure of an invention. This delay could be risky because priority goes to the first applicant to file. This "first-to-file" rule is much easier to adjudicate than the "first-to-invent" rule, used in some other countries, which can involve litigation to decide who was the first inventor.

Patent applications may take years to process, but they are made public 18 months after filing. Protests may be filed with the Patent Office to challenge the application. For example, an attempt to patent videoconferencing was successfully challenged.

An invention may be legally made by others, used, or sold in the period after publication but before issuance of the patent; however, the inventor can seek retroactive compensation when the patent is awarded. Many articles are stamped "patent pending" to warn that retroactive compensation may be sought.

In order to establish a filing date, at least a petition, a defining description, and a fee must be submitted. The full patent application must be completed within 15 months and must contain the following five parts:

- petition: a formal request that a patent be granted,
- abstract: a summary for publication in the Patent Office Record,
- specification: a description of the invention, including relevant prior art, and its novelty, usefulness, and ingenuity,
- claims: a part of the specification that defines precisely what aspects of the invention the inventor wants to protect,
- drawings: illustrations (and possibly models) of the invention.

The most important part of the application is the scope of the claims. These define the protection provided by the patent and are usually as broad as possible to protect against competitors who produce similar inventions with minor differences.

Once the patent application has been filed, a request must be made to have it examined. Such a request must be made within five years of the filing date; otherwise the application expires. During this delay, the applicant may be engaged in market research and business planning to see if it is worthwhile to proceed.

There are fees for filing, for the examination, for the granting of the patent, and annual maintenance fees. Protecting the patent against infringement entails legal fees, so a patent can be an expensive undertaking in total.

Canadian patents are valid only in Canada, and applications must be filed in each country where protection is desired. However, the international Paris Convention for Patents (1883) makes patent applications in member countries valid as of the date granted in the home country, provided application is made abroad within a year. Most industrialized countries participate in the Patent Cooperation Treaty of the World Intellectual Property Organization in Geneva, which permits a foreign patent to be obtained from within Canada. The standardized filing procedure simplifies the application, so you can apply to over 100 countries with one application in Canada.

The Patent Office cannot assist inventors to lease or sell patents; however, a patent owner may place an advertisement in the Patent Office Record, which is published weekly and is available in any large Canadian public library. Many entrepreneurs, investors, and researchers consult this publication on a regular basis.

4 Copyright

According to the Copyright Act as administered by the CIPO [7], a copyright is the right to produce, reproduce, perform, publish, adapt, sell, or lease an original literary, dramatic, musical, or artistic work. Reproduction includes recording, photographing, filming, or using any "mechanical contrivance," including communications technology and telecommunication signals, for delivering the subject matter.

The author owns the rights to an original work, unless the work was created as part of a job or employment or is a commissioned work. If the creation occurs during an employment, the employer usually owns the rights, unless there is an agreement to the contrary.

Protection of a copyright lasts for the author's lifetime plus 50 years. Photographs, sound recordings, and most films and videos are protected for 50 years from the date of their creation.

Material subject to copyright may be printed or written but also may be recorded on electronic media. Computer programs, technical reports, and engineering drawings and data are considered to be literary or artistic works and are protected by copyright.

4.1 Copyright registration

In Canada, a copyright is automatically possessed by an author or creator, whether or not, as required in some other countries, the material is marked "©" or is registered.

Registration is not necessary to obtain copyright protection. However, you may still want to register the copyright with the CIPO for a small, one-time fee. Registration establishes the date of creation, the original author's name, the ownership of the work, and gives notice that a copyright exists.

Canadian copyright also gives foreign protection as a result of two international treaties. Automatic ownership of copyright is valid in all countries that adhere to the Berne Copyright Convention. Furthermore, copyright can be extended to all countries that adhere to the Universal Copyright Convention by marking the creation with the copyright symbol ©, the name of the author, and the year. Canada and the United States belong to both conventions.

Violation of a copyright is called infringement, and if it includes misrepresentation of authorship then it is plagiarism, as discussed in Section 1.2 of Chapter 8 and Section 4 of Chapter 3. Simple infringement occurs with a substantial quotation or borrowing of another's work without the author's permission, even if the source is attributed properly [3].

Enforcing a copyright is the owner's responsibility. If someone infringes your copyright, you may sue under civil law for the recovery of lost income. However, the Copyright Act also has clauses for criminal infringement when serious, organized infringement or "piracy" occurs, such as reproducing a copyrighted work for sale or hire. The penalty for criminal infringement is severe.

4.2 Fair dealing

The Copyright Act permits a small amount of copying or reproduction under "fair dealing" provisions. These include short extracts for the purpose of review, criticism, or research. The author and the source of the quoted material must be completely identified. However, the amount that can be copied is not well defined. If copying replaces sales, it is certainly not "fair." Multiple copying always requires the consent of the copyright owner.

Photocopying written work for research and study is allowed but controversial, since there are no exact guidelines for the number of paragraphs or pages that can be copied under fair dealing. Obviously, photocopying an entire work, such as a textbook, is illegal infringement. Universities, therefore, have very strict rules about copying. Licensing collectives, such as CANCOPY, have been established to negotiate licences for schools and universities and to pay royalties to authors.

4.3 Copyright and computer programs

Computer programs may be protected by copyright. A program can be copyrighted as a literary work, regardless of the manner of storage or recording. Under the Copyright Act, it is an infringement to make, sell, distribute, import, or rent copies of computer programs, without written consent of the copyright owner. As noted in Section 3.1, a process with a computer program embedded in it may be patentable, but the written program itself is a literary work and may be protected by copyright.

Fair dealing, discussed in Section 4.2, also applies to computer programs. The owner of an authorized copy may make one back-up copy for his or her own use. This copy must be destroyed when the authorization ends for the original. The authorized copy may also be adapted or translated into another computer language, if this is necessary to make it compatible with a particular computer system.

The "piracy," or unauthorized copying of software, is so easily done that people sometimes do not realize that they are breaking the Copyright Act. As a professional engineer, you must avoid unethical activities that might cloud your reputation. Avoid pirated software, and ensure that all your computer software is authorized.

The interpretation of copyright law for computer programs is still developing. For example, a form of reverse engineering, called *program decompilation*, may be much cheaper than obtaining a licence or developing a completely new program, but the legal issue is whether decompilation represents fair dealing. A single decompiled copy may be considered fair dealing, if decompilation is used merely to inspect competing software.

Most software user manuals claim copyright on the manual and on the software itself. The medium on which the software is stored may be under warranty, but not the performance of the software. The intellectual property is not sold; it is only licensed for use by the purchaser, who is permitted one back-up copy.

5 Industrial designs

An industrial design is any original shape, pattern, or ornamentation applied to a useful article of manufacture. Registration of an industrial design gives the owner exclusive rights for five years, renewable once for a further five years. Registration protects the identity of the owner in the marketplace, where the visual appeal of a product may distinguish it from other products with similar function. Some examples are the decorative pattern on the handles of a cutlery set, a wallpaper pattern, or an unusual but appealing chair shape. Industrial design protection is like copyright but with some differences, as follows:

- Industrial design protection applies to the shape, pattern, or ornamentation of mass-produced, useful articles. The visual aspects of the design are protected.
- Industrial design protection can be obtained only by registration, and the article must be registered during the first year of public or published use.
- The protection has a maximum term of 10 years.
- If the design or pattern is on an article or piece of work that is subject to copyright but is used to produce 50 or more such articles, then it can only be protected as an industrial design.

The Industrial Design Act regulates the registration of industrial designs [8]. The application for registration is made by the proprietor of the design and must contain four basic parts:

- a written description of the features of the design that are original,
- a depiction (drawings or photographs) of the design,
- a declaration that you are the proprietor and that, to your knowledge, it is original; that is, no one has used the design before,
- the registration fee.

A product does not have to be marked, but marking the label or package of the product with the letter D inside a circle, with the name of the proprietor, increases the degree of protection.

Industrial design protection applies in Canada only. Separate registration must be made for foreign countries, but under the Paris Convention, a person who applies for protection in Canada has six months to apply for registration in other treaty countries.

6 Trademarks

A trademark is a word, symbol, or design intended to identify the wares or services of a person or company in the marketplace [9]. The following types are included:

- *Ordinary marks* distinguish articles or services. Most corporate logos fall into this category.
- *Certification marks* indicate a quality standard for a product or service set by some recognized organization (see Figure 17.3).
- *Distinguishing guise* is a wrapping, shape, packaging, or appearance that serves to identify the product as distinct from other similar products or services. An example is a candy shaped like a butterfly.

Fig. 17.3. Three examples of registered certification marks from the CIPO database are shown. The first is a Canadian Automobile Association mark familiar to consumers. The second is registered by the Air-conditioning and Refrigeration Institute, and the third is used by members of the Association of Professional Engineers and Geoscientists of New Brunswick (with permission of the trademark holders).

The trademark owner benefits from public recognition of the quality and reputation of the product or service. A company name does not necessarily qualify as a trademark unless it is used to identify products or services. A trade name or business name used to identify products or services can be registered as a trademark.

The right to register a mark in Canada belongs to the first person to use it in Canada or to a person with a registered trademark in a member country of the Union for the Protection of Industrial Property. The term of registration is 15 years from the date of original registration or from the most recent renewal. To be continued in effect, renewal must take place before expiry, but the number of terms of renewal is unlimited. While an unregistered trademark may be commonly recognized, only registration in the trademark register is proof of exclusive ownership and protection against infringement. The process of registration also ensures that the proposed mark does not infringe existing trademarks.

A proposed trademark is deemed to be registerable unless it falls into one of the following categories:

- It is similar to an existing registered trademark.

- It is the applicant's name or surname unless the name has become associated with a product and it identifies that product.
- It is a clearly descriptive term. For example, "Red" cherries cannot be registered, since all cherries are red, and it would be unfair to other cherry sellers.
- It is a deceptively incorrect term. For example, "Air Courier" is not allowed for a company that delivers parcels by truck.
- It is the name of the product in English or any other language. For example, "Fleurs" could not be registered for a flower shop.
- It suggests government approval; for example, a mark including the Canadian flag.
- It suggests a connection with the Red Cross, Armed Forces, a university, public authority, or the RCMP.
- It could be confused with international signs, such as traffic or civil defence.
- It could be confused with a plant variety designation.
- It is the name of a living person or one who has died in the preceding 30 years, unless the person has given written consent.

The Canadian Intellectual Property Office ensures that no two trademarks are the same or confusingly similar.

It is recommended, although not legally necessary, that registered trademarks be identified by the symbols ® or ™ or *Trademark Registered, or ᴹᶜ or Marque de Commerce. Trademark owners must be vigilant and point out any improper use, such as being copied or used as a generic or common name. To reduce this risk, the trademark should always be used as an adjective, never as a noun. For example, kerosene and cellophane are generic nouns derived from former trademarks for Kerosene fuel and Cellophane sheet.

Registering a trademark protects it only in Canada. For countries in the Union for the Protection of Industrial Property, registration is obtained by filing a copy of the Canadian registration. For a foreign trademark to be registered in Canada, it must be distinctive, non-confusing, non-deceptive, and not otherwise prohibited.

7 Integrated circuit topographies

Integrated circuit topographies are defined under the Integrated Circuit Topography Act (1993) [10]. The intellectual property is the three-dimensional configuration of layers of semi-conductors, metals, insulators, and other materials that form the circuits in the semi-conductor microchip. To be registered, the topography must be original. Registration of the topography can supplement patent protection for a novel circuit itself or its function.

This type of protection is very new, and the laws will undoubtedly change as time passes. Over 20 countries have similar protection for semi-conductor chips.

Fair dealing provisions allow copying the design for research and teaching and limited reverse engineering.

Registration is required for protection, must be applied for within two years of commercial use of the design, and is valid only in Canada.

The term is to the end of the tenth year from the first commercial use or from the filing of the application for registration, whichever is earlier.

8 Trade secrets

Trade secrets are intellectual property; however, the government does not regulate them. A trade secret is a commercially important secret formula, design, process, or compilation of information. The purpose of maintaining secrecy is evident: to retain exclusive use and benefit from a process or product. The benefits may be great, the term of use is potentially indefinite, and there are no fees. However, this form of protection works only as long as the secret can be maintained.

Unlike other forms of intellectual property, there is no specific government act that protects trade secrets. The other forms of protection inherently require disclosure as a condition of protection, which provides a source of information and a stimulus to technological progress.

The action required to protect a trade secret is simply to keep all the relevant information secret. Employees who have access must agree to maintain secrecy, usually as part of their employment contract. If a trade secret is deliberately or accidentally misappropriated, then legal action may be undertaken for damages, but requires recourse to the laws for theft or tort (a willfully or negligently wrongful act).

Trade secrets are subject to leaks, espionage, independent discovery, reverse engineering, or obsolescence, and the cost of security to maintain secrecy may be very high. Often, the main problems are to control carelessness and to ensure the confidentiality of information held by former employees, perhaps now employed by a competitor!

9 Further study

1. What type, or types, of intellectual property can be used to protect the following?

(a) A new pneumatic pump that makes athletic shoes more comfortable.

(b) The attractive design of a new pair of athletic shoes.

(c) A logo to identify the new line of athletic shoes.

(d) A slogan to advertise the new line of athletic shoes on television.

(e) A new process for curing rubber used in athletic shoes.

(f) A new microcomputer chip to control the production line for athletic shoes.

2. Engineering reports were discussed in Chapter 8. Compare the headings in a patent application with the headings in a typical engineering report. What are the similarities and differences?

3. Omar Khayyám, an eminent Persian astronomer who died in 1123, wrote many verses in the Persian language. These were collected and translated into poetic English in 1858 as The Rubáiyát of Omar Khayyám by Edward FitzGerald (1809–83). Imagine that current Canadian copyright law had applied for five centuries in Persia and England.

   (a) When would Khayyám's copyright on his work have expired?
   (b) Could FitzGerald copyright the translation in 1858?
   (c) If the answer to (b) is affirmative, for how long would FitzGerald's copyright be valid?
   (d) If you publish a newly illustrated version of FitzGerald's Rubáiyát, can you copyright it in the year 2002?

4. While working for the Grand Whiz Company, you imagine a new product. It is not directly related to the projects you have worked on there, but there is no doubt that the working environment, and especially the discussions with your working colleagues, has stimulated your imagination.

   Do you: (a) take the idea to the Grand Whiz management for a shared patent, if the company bears all the expense of the patent process; (b) quit, and then set up your own business to exploit your idea; (c) keep your job, and take the idea to your entrepreneurial sister-in-law and share the profits with her?

5. Competitors often seek proprietary business information. If the methods are legal, they are called market research; if the methods are questionable, they are called industrial espionage. One of the easiest ways to acquire intellectual property from a competitor is to hire someone who has worked there.

   (a) What are some ways of preventing the loss of intellectual property in this way?
   (b) What is the career effect for an employee who easily transfers jobs for this reason?

6. How would you protect the following items of intellectual property?

   (a) a company name and logo,
   (b) a story, poem, or laboratory report,
   (c) a natural law, if you discovered it,

(d) a distinctive design for a mass-produced item,

(e) a novel and ingenious mousetrap.

Answers:
1. (a) The function of a new pneumatic pump can be protected by patent. (b) Shoe design can be registered as an industrial design, if it is original. (c) A logo that identifies a product (shoes) may be registered as a trademark. (d) A mere slogan cannot be protected; however, a videotaped commercial containing the slogan is protected under copyright. (e) A new process for curing rubber can be protected under patent; it might also be possible to keep it a trade secret. (f) The circuit layout of the microcomputer chip can be protected as integrated circuit topography.
3. (a) 1173; (b) Yes; (c) 1933; (d) Yes.
5. (a) Non-disclosure agreements for proprietary information, fair remuneration, reference checks, computer access controls. (b) Trail of poor references; bad reputation on ethics and confidentiality.
6. (a) Trademark; (b) Copyright; (c) Cannot be protected; (d) Industrial design; (e) Patent.

10 References

[1] Harry Goldstein, "The secret art of chip graffiti," *IEEE Spectrum*, vol. 39, no. 3, pp. 50–55, 2002.

[2] Canadian Intellectual Property Office, *Canadian Intellectual Property Office—Welcome*, Industry Canada, Canadian Intellectual Property Office, 2001, <http://strategis.ic.gc.ca/sc_mrksv/cipo/welcome/welcom-e.html> (April 6, 2002).

[3] W. Hagen, *IEEE Intellectual Property Rights*, Institute of Electrical and Electronics Engineers, New York, 2001, <http://www.ieee.org/about/documentation/copyright/> (March 14, 2002).

[4] Canadian Intellectual Property Office, *A Guide to Patents*, Industry Canada, Canadian Intellectual Property Office, Ottawa, 2001, <http://strategis.gc.ca/sc_mrksv/cipo/patents/pat_gd_main-e.html> (March 14, 2002).

[5] B. Bunch and A. Hellemans, *The Timetables of Technology*, Simon & Schuster Inc., New York, 1993.

[6] L. Brillouin, *Science and Information Theory*, Academic Press, New York, 1962.

[7] Canadian Intellectual Property Office, *A Guide to Copyrights*, Industry Canada, Canadian Intellectual Property Office, Ottawa, 2000, <http://strategis.gc.ca/sc_mrksv/cipo/cp/copy_gd_main-e.html> (March 14, 2002).

[8] Canadian Intellectual Property Office, *A Guide to Industrial Designs*, Industry Canada, Canadian Intellectual Property Office, Ottawa, 2001, <http://strategis.gc.ca/sc_mrksv/cipo/id/id_gd_main-e.html> (March 14, 2002).

[9] Canadian Intellectual Property Office, *A Guide to Trade-Marks*, Industry Canada, Canadian Intellectual Property Office, Ottawa, 2001, <http://strategis.gc.ca/sc_mrksv/cipo/tm/tm_gd_main-e.html> (March 14, 2002).

[10] Canadian Intellectual Property Office, *A Guide to Integrated Circuit Topographies*, Industry Canada, Canadian Intellectual Property Office, Ottawa, 2001, <http://strategis.ic.gc.ca/sc_mrksv/cipo/ict/ict_gd_main-e.html> (March 14, 2002).

Chapter 18

Project planning and scheduling

Planning and scheduling are important parts of any engineering project, whether it involves design, development, or construction. In fact, almost every aspect of everyday life could be improved by better planning or scheduling. These two basic terms can be defined as follows:

- *Planning* is the determination of all the activities involved in a project or similar enterprise and their arrangement in a logical order.
- *Scheduling* is the assignment of beginning and end times to the activities.

Planning, then, is determining *how* to do something, and scheduling is deciding *when* to do it. In any project, the purpose of planning and scheduling is to achieve minimum time, minimum waste of time and material, and minimum total cost. These objectives cannot always be achieved. For example, achieving minimum time is typically difficult at minimum cost. However, with proper planning and scheduling, the trade-off between these objectives can be seen in advance and optimal choices can be made. The material in this chapter is contained in references such as [1–4].

1 Gantt charts and the critical-path method

Charts can be used to aid the visualization of planning and scheduling activities. There are several suitable formats, but the most common is probably the bar chart, also called the *Gantt chart,* after its inventor. Many project-management software tools assist with the creation and modification of such charts. A basic example is shown in Figure 18.1. The chart shows activities, in order of starting times, on the vertical scale; the horizontal scale is time. The time intervals of the activities are indicated by horizontal bars, and as the project proceeds, notations may be added or the style of the bars modified to show activities, for example, that are critical or exceeding their planned duration. At any time, the activities in progress and those scheduled to start or end soon can be determined. However, if a problem occurs with an activity (e.g., if delivery of material is late), the basic Gantt chart does not show whether other activities will be affected or whether the delay will affect the completion date of the whole project.

A technique called the *critical-path method* (CPM) gives detailed scheduling information. CPM is a modified form of the "program evaluation and review technique" (PERT), developed by the U.S. Navy in 1958 to speed the design of the

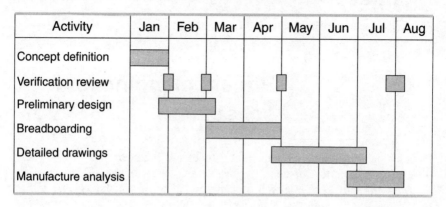

Fig. 18.1. A basic Gantt chart for a design project.

Polaris missile. It proved to be an instant success by cutting 18 months from the completion date of the project, and it has been the standard method of planning used in aerospace and electronic industries and in many construction projects ever since. CPM is particularly suitable for projects composed of easily defined activities of which the durations can be accurately estimated (most construction projects and some software-engineering projects, for example, fall into this category).

Many planners insist upon CPM being used in controlling a project, but they still put information in the simple form of Gantt charts when discussing or reporting on the project.

2 Planning with CPM

To plan a project using CPM, follow these two steps:

1. List each activity involved in the completion of the project.
2. Construct an arrow diagram that shows the logical order of the activities, one activity per arrow.

When the arrow diagram is complete, it shows all the activities in the project, in the proper order. The planning phase is therefore complete. The two steps above may actually be repeated several times, since the action of constructing the arrow diagram will remind the planner that an activity has been omitted or perhaps that one activity should be divided into two more specific activities. The list of activities and the arrow diagram are then modified until they are complete and correct. However, before discussing details of the arrow diagram, we must define two terms more accurately: activities and events.

- An *activity* is any defined job or process that is an essential part of the project. Each activity in the project is represented by an arrow in the arrow diagram such as shown in Figure 18.2. Each activity has an associated time duration Δt.

- An *event* is a defined time. All activities begin and end at events, represented by circles or nodes, as shown in Figure 18.2. In some contexts, events are called "milestones," which are defined steps toward completion of the project.

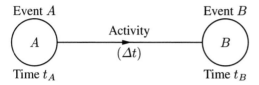

Fig. 18.2. Activities and events in an arrow diagram, with time interval $\Delta t = t_B - t_A$.

An arrow diagram, therefore, is a graph, in which the nodes are events and the branches are arrows representing activities. To construct an arrow diagram, we start by drawing a circle for the *start event*, which is the beginning of the project. The first activity arrow is then drawn from the start event and the other activity arrows are joined to it in the sequence that the activities must be performed. The following simple example illustrates the planning procedure.

Example 1. Tire changing.

For three or four weeks each autumn, when the weather gets cold, tire stores (in some climatic regions) are crowded with people who want to purchase winter tires. If these stores could improve the efficiency of selling and mounting tires, then there would be fewer delays, customers would be more content, and tire sales would increase.

The activities for a typical tire purchase are listed in Table 18.1 along with estimated activity times. Consider a typical tire purchase as a project. How would you plan and schedule it for minimum delay? Consider two cases: a small, one-person tire store and a large, fully staffed tire store.

The arrow diagram can be created easily, using the list of activities in the table. If the tire store has only one person acting as sales clerk, stock-room clerk, and mechanic, then the simple diagram of Figure 18.3 results. The process cannot be improved if only one person is involved, since the activities must be performed sequentially.

However, if the tire store employs a sales clerk, a stock-room clerk, and *two* mechanics (and has two tire-mounting machines), then the arrow diagram can be re-drawn as shown in Figure 18.4. Since both tires are installed at the same time, the time for the transaction is reduced, as will be seen.

We now look at the process critically: although the list of activities is in a logical order, beginning with the arrival of the customer and ending with the departure, it is not necessarily the most efficient order. For example, why must the

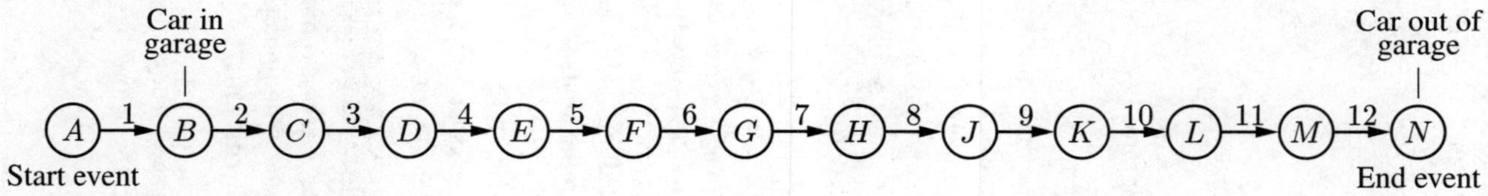

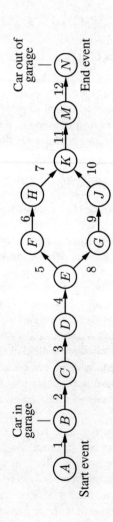

Fig. 18.3. Arrow diagram for a one-operator tire purchase.

Fig. 18.4. Arrow diagram for a large-store tire purchase.

Table 18.1. Activities for the sale and installation of winter tires.

	Activity	Time (min)
1.	Customer drives car into tire-store garage	2
2.	Customer inspects catalogue and selects tires	30
3.	Sales clerk confirms stock on hand	5
4.	Stock-room clerk takes tires to garage	10
5.	Mechanic raises car and removes wheel 1	5
6.	Mechanic mounts new tire on wheel 1 and balances it	10
7.	Mechanic replaces wheel 1	5
8.	Mechanic removes wheel 2	3
9.	Mechanic mounts new tire on wheel 2 and balances it	10
10.	Mechanic replaces wheel 2	5
11.	Clerk writes bill; client pays for tires and installation	10
12.	Mechanic lowers car and drives it out of garage	2

car be in the garage while the customer inspects the catalogue or pays the bill? The bill could be paid either before or after the tire change, thus freeing the garage for part of the time. With these changes in mind, the process can be re-drawn as in Figure 18.5, showing that the customer parks the car, inspects the catalogue (activity 2), selects the tires (3), and pays the bill (11) before the car enters the garage (1). The stock-clerk delivers the tires to the garage (4) while the customer pays and drives the car into the garage. The dashed lines in this figure are called *dummy activities,* which take zero time but are included to show the proper sequence of events. In this case, they are required in order to show that the tires must be in the garage (4) for mounting to take place (6, 9). In the next section, the concept of scheduling is introduced, and we can calculate the amount of time saved.

3 Scheduling with CPM

When the initial planning is complete, scheduling can begin. We say "initial" planning because we may, as a result of scheduling, decide to change the plan. Two terms associated with scheduling must be defined:

- The *earliest event time* (EE) for an event is the earliest time at which the activities that precede the event can be completed.

226 Chapter 18 Project planning and scheduling

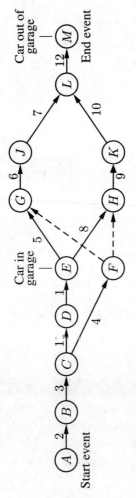

Fig. 18.5. Arrow diagram for the improved procedure.

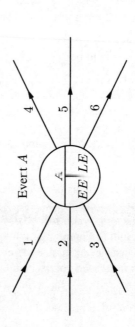

Fig. 18.6. Format for writing earliest event time (EE) and latest event time (LE) in an event circle.

- The *latest event time* (LE) for an event is the latest time at which the activities that follow the event can commence without delaying the project.

The EE and LE times are easily calculated, as shown on the following pages. Since the EE and LE times are associated with events, they are usually included in the event circles (the nodes of the graph), as shown in Figure 18.6, which help to make the arrow diagram more understandable.

Calculating earliest event times: The EE value at the start event is zero. At any other event, the EE value is the sum of the activity times for the arrows from the start event to that event, but if there is more than one path from the start event, the EE value is the largest of these path times. The calculation is usually done by working from left to right until the end event is reached. The EE value at the end event is the total time T required for the project.

Calculating latest event times: The calculation of LE times is similar to that of EE times, except that the path traversal is carried out in reverse, beginning with the end event. The end event LE value is T. To calculate the LE value for any other event, the activity times for the arrows between that event and the end event are subtracted from the total time T, but if there are two or more paths to the end event, the largest of the path times is subtracted from T. The LE value for the start event must be zero, which serves as a check for arithmetical errors.

The critical path: A path with the longest path time from the start event to the end event is said to be "critical," since additional delay in such a path delays the whole project. Conversely, if the critical path time is reduced, then a reduction of the project time may be possible. All non-critical paths have spare time available for some activities. Events with equal EE and LE values are on a critical path.

Example 2. Tire changing: Example 1 (continued).

Consider again the tire-changing example discussed previously; the list of activities and activity times is shown in Table 18.1. We want to find the critical path for both the one-person and fully staffed stores.

The planning graph for the one-person operation, shown in Figure 18.3, has been re-drawn in Figure 18.7, showing EE and LE values in the event circles. The activity time Δt is shown in parentheses below each arrow. As discussed previously for the one-person operation, the CPM arrow diagram is a simple sequence of activities, so the relationships between Δt, EE, LE, and T values are simple. The total project time is the sum of the activity times; that is, $T = 97$ min. In this simple case, $EE = LE$ at every event, so every activity and every event is on the critical path. If there is a change in any activity time, there is a corresponding change in the project completion time.

Consider the fully staffed tire store of Figure 18.8, which is Figure 18.4 with the scheduling information included. The figure shows that by mounting two tires (activities 6, 9) at the same time, it is possible to reduce the total time of the project to $T = 79$ min, as discussed earlier. The critical path is indicated in Figure 18.8 by thick arrows. To reduce the total project time, it is necessary to examine and improve activities on this path.

The improved procedure, discussed earlier and illustrated in Figure 18.5 is redrawn in Figure 18.9 with the scheduling information included. The total project time has been further reduced: now $T = 69$ min. However, by grouping all the sales and billing activities and separating them from the garage activities, it is seen that the car is now in the garage from $EE = 47$ to $EE = 69$, a total of $69 - 47 = 22$ min. This is clearly an improvement, freeing the garage for other work. The project could probably be further improved, but this single cycle of planning and scheduling has illustrated how a methodical approach leads the planner, almost automatically, to think about methods of improvement.

A final important definition will be introduced. Activities that are not on the critical path have some spare time, usually called *float time* or *slack time*. We can calculate this time F as follows: If an activity starts at event A and ends at event B, then F equals LE at event B minus EE at event A, minus the activity time Δt. In equation form:

$$F = LE_B - EE_A - \Delta t. \tag{18.1}$$

As an example, consider activity 9 in Figure 18.9, which can start as early as $EE = 50$ (event H) and must end before $LE = 62$ (event K). The difference is 12 min, although activity 9 requires only 10 min. Therefore we have $62 - 50 - 10 = 2$ min float time.

The float time is significant for project managers. If a delay occurs during an activity, the completion of project will not be affected, as long as the delay is less than the float time for that activity. Activities on the critical path have zero float time.

4 Refinement of CPM

The tire-changing example of Section 3 is rather simple, but it includes all of the basic CPM concepts. It shows how the CPM method forces the planner to ask questions that may lead to improvements. The improvement of an adequate but inefficient plan is called "optimization." In simple cases, the list of activities, shows the planner the optimum course of action and the arrow diagram may not be needed.

It may be thought that the techniques for calculating EE, LE times and the total time T are rather formal and mechanical. However, the calculation of the total project time T can be done in no other way, regardless of the scheduling technique. The reverse process, the calculation of LE values and the identification of the critical path, is unique to CPM and distinguishes it from other methods. The critical path and the LE times are valuable when delays occur in the middle of a project. The use of CPM facilitates decisions that may counteract or eliminate the delays.

Section 4 Refinement of CPM 229

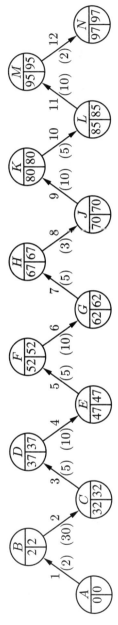

Fig. 18.7. Arrow diagram for one-operator, from Figure 18.3, showing event times and a total time $T = 97$ min.

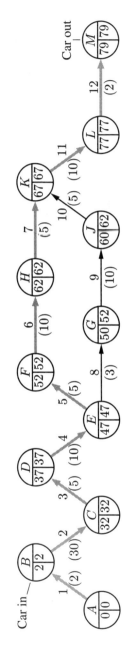

Fig. 18.8. Arrow diagram for the large store, from Figure 18.4, showing event times and a total time $T = 79$ min.

The arrow diagrams sometimes appear quite complicated, even for simple problems such as the tire-store example discussed above. However, the diagram is a permanent record of the logical decisions made in the planning process. The diagram becomes more important as the number of activities increases. When the activity count reaches about 100, it is necessary to use one of the many CPM computer programs developed to calculate EE, LE, T, and float times. These programs are usually arranged to list the activities in order by start times; some will produce a Gantt chart for a project, which is usually easy to understand. The Gantt chart for the tire-store example is shown in Figure 18.10, where the solid bars are activity times and the gray bars are float time. The chart in Figure 18.10 agrees with the arrow diagram in Figure 18.9.

5 Summary of steps in CPM

The following eight steps outline the use of CPM in planning and scheduling a project:

1. List all the activities in the project, and estimate the time required for each activity.
2. Construct the arrow diagram, in which each arrow represents an activity and each event (node) represents a point in time.
3. Calculate the earliest event time (EE) for each event by calculating the maximum of the path times from the start event. The end event EE value is the total project time T.
4. Calculate the latest event times (LE) for each event by subtracting the maximum of the path times to the end event from T. The start event LE should be zero.
5. Calculate the float time for each activity using Equation (18.1).
6. Identify the critical path. Events that have identical EE and LE values have no spare time and are on the critical path. The activities that join these events and have zero float time are the critical path activities.
7. Optimize the project. When the diagram is finished and values of EE, LE, T, and F have been calculated, it is possible to survey the project and look for ways to reduce time and cost, eliminate problems, and prevent bottlenecks.
8. Use the information obtained to keep the project under control. Do not use it just for initial planning and file it away. The main strength of CPM is its ability to identify courses of action when crises occur during the project.

Section 5 Summary of steps in CPM 231

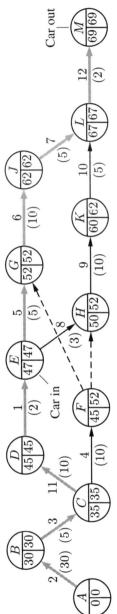

Fig. 18.9. The improved procedure, from Figure 18.5, showing event times and a total time $T = 69$ min.

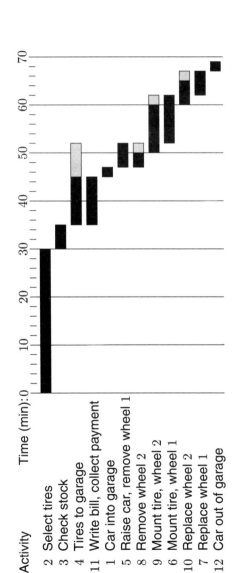

Fig. 18.10. Gantt chart for the tire-changing example, showing the improved procedure of Figure 18.9 with slack times.

Activity	
2	Select tires
3	Check stock
4	Tires to garage
11	Write bill, collect payment
1	Car into garage
5	Raise car, remove wheel 1
8	Remove wheel 2
9	Mount tire, wheel 2
6	Mount tire, wheel 1
10	Replace wheel 2
7	Replace wheel 1
12	Car out of garage

6 References

[1] K. G. Lockyer and J. Gordon, *Project Management and Project Network Techniques*, Pitman Publishing, London, 6th edition, 1996.

[2] J. J. O'Brien and F. L. Plotnick, *CPM in Construction Management*, McGraw-Hill, New York, 5th edition, 1999.

[3] A. Harrison, *A Survival Guide to Critical Path Analysis*, Butterworth-Heinemann, Oxford, 1997.

[4] S. A. Devaux, *Total Project Control: A Manager's Guide to Integrated Project Planning, Measuring, and Tracking*, John Wiley & Sons, Inc., New York, 1999.

Chapter 19

Safety in engineering design

No engineering design or workplace environment can ever be absolutely free of hazards; however, usually with little effort it is possible to reduce the risk of injury or illness to an acceptable level. Although all engineers must be concerned with safety, these matters can also be considered as a technical specialty, and there is a technical society [1] for loss prevention specialists.

To deal with hazardous situations effectively, the engineer needs a methodical procedure for recognizing hazards and introducing remedies. This chapter discusses some basic methods of recognizing hazards and reducing them by introducing

- safety responsibilities of the engineer,
- guidelines and principles for recognizing and controlling hazards,
- applicable safety codes and standards.

1 Responsibility of the design engineer

Health and safety issues usually receive public attention only after an accident such as a bridge collapse, a railway derailment, a spill of toxic chemicals, or a gas explosion. These disasters are reported in the news from time to time, although the numbers of such incidents have decreased over the years as education, codes, and standards have improved. For example, in the late 1800s, when steam was first being used for engines and heating, boiler explosions, causing fatalities through scalding, were fairly common. However, since the development of the ASME boiler and pressure-vessel design code, boiler explosions have disappeared, even though steam at much higher temperatures and pressures is widely used for electrical power generation and for heating large buildings.

Engineers may encounter situations that have the potential to be dangerous. Examples of such situations are the presence of high voltages, high temperatures, high velocities, toxic chemicals, and large amounts of energy, whether pneumatic, hydraulic, thermal, electrical, or other kinds. In addition, during construction and manufacturing processes, hazards may arise in the factory, worksite, or field installation that can create health or safety problems for those present.

An engineer carries a serious responsibility when supervising a design or construction project, because failure to act to correct a potentially dangerous situation or failure to follow codes and standards is, by definition, professional misconduct.

The engineer who neglects health and safety aspects of a design runs the double risk of disciplinary action by Professional Engineers Ontario (PEO), as well as potential legal liability if damage or injury should occur as a result of design deficiencies or inadequate safety measures.

There are two groups of people whose health or safety must be considered: the eventual users of a device, product, or structure being designed and the workers involved in the manufacture or fabrication of it. The safety of any design must be examined with respect to both of these aspects.

2 Some general guidelines

There is a general procedure, described briefly below and explained in more detail later in this chapter, for dealing with a hazard:

1. Identify the hazard. Table 19.1 may help in hazard identification.
2. Try to prevent or eliminate the need for creating the hazard.
3. If the hazard cannot be eliminated, then it should be treated as a form of signal that emanates from a "source" and follows some "path" to a "receiver," the human worker or user of the design, where it may inflict some damage. This analogy then identifies three locations where action can be taken to prevent the damage: at the source, along the path, and at the receiver.
4. Finally, if the above steps prove unsuccessful and the resulting design proves to be unsafe, then a remedial action is essential: recall the unsafe devices, notify people of danger, assist the injured, and so on, as appropriate.

When hazards cannot be eliminated, the signs recommended by standards institutes, for example, the American National Standards Institute (ANSI), should be used. Three common examples are illustrated in Figure 19.1. Signs of the same type should be of consistent design in order to avoid confusion. In addition, the wording should be clear and concise. The three examples from Figure 19.1 are briefly described below:

CAUTION: used to warn of risks that might result from unsafe practices. The sign should have a yellow background. The word Caution is yellow on a black upper panel.

WARNING: denotes a specific potential hazard. The warning sign usually has an orange background. The word Warning is orange on a black upper panel.

DANGER: indicates that a most serious potential hazard to personal safety is in the vicinity of the sign. The word Danger should be in white letters on a red oval in a black rectangle.

Section 2 Some general guidelines 235

Table 19.1. Examples of hazard control methods.

HAZARD	CONTROL MEASURES		
	SOURCE	PATH	RECEIVER
MECHANICAL	1. Enclosure guards 2. Interlocking guards (mechanical, electrical) 3. Reduction in speed 4. Limitations of movement	1. Guard by location (rope off area, etc.) 2. Remote control	1. Education 2. Rules and regulations for clothing, etc. 3. Pull-away devices 4. Aids for placing, feeding, ejecting workpieces 5. Two-hand trip switch buttons
NOISE	1. Enclosure 2. Surface treatment 3. Reduction of impact forces	1. Building layout 2. Increase distance 3. Acoustic fibres 4. Mufflers	1. Protective equipment: earmuffs, earplugs 2. Limit exposure time
ELECTRICAL	1. Low voltage instruments 2. Fuses, circuit breakers 3. Insulation 4. Lock-outs 5. Labelling and test points	1. Grounding 2. Use of ground fault detectors	1. Protective equipment 2. Education
THERMAL (HEAT OR COLD)	1. Shielding 2. Insulation 3. Painting 4. Ventilation 5. Limiting the physical demands of the job	1. General ventilation	1. Select acclimatized personnel 2. Acclimatization program 3. Adequate supply of water 4. Special clothing; ventilated suits 5. Proper work-rest schedules 6. Limiting exposure time
CHEMICAL	1. Isolate or substitute 2. Change of process	1. Ventilation	1. Protective equipment, respirators, etc.

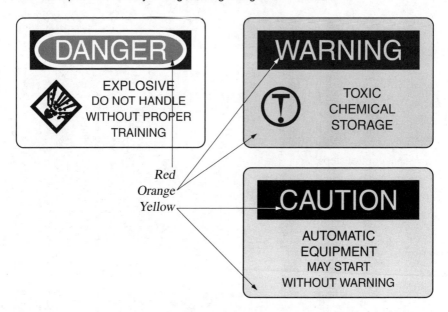

Fig. 19.1. Examples of standard hazard signs are shown. The colour and upper panel are specified for each type; specific information is placed in the bottom region.

3 Principles of hazard recognition and control

The following is an outline of a basic sequence of steps for assessing hazards.

1. **Assess the capabilities and limitations of users.** Designers must realize that the equipment or products they are creating will ultimately be used by human beings who vary considerably in both physical and mental abilities. Designing an item for a general user is more demanding than creating products for highly select user groups, such as airline pilots. General users have wide variations in abilities, and their minimum skill level should not be overrated.

2. **Anticipate common errors and modes of failure.** Experience has shown that when a design goes into widespread use, there is validity to Murphy's law: "Anything that can go wrong will go wrong." Thus designers must try to anticipate what can go wrong with their creations and then modify the design to prevent these unwanted events, or if this is not possible, at least to minimize their consequences and warn of the hazard.

   How can this be done? In general, it means asking a series of questions such as: Can a given event or sequence take place? If it can, what will be the result? Can this result lead to illness, injury, or death? If so, what can be done to prevent this? This is known as "idiot proofing."

3. **Evaluate designs for safety and health.** Evaluation for safety and health risks is usually known as "design review." While this process is by no means foolproof, it can help considerably in the elimination of most obvious hazards. It may also prove helpful in aiding a manufacturer's defence against a future product liability lawsuit by demonstrating that he or she was conscious of the hazardous nature of the product and undertook a series of prudent control steps.

4. **Provide adequate instructions and warnings.** Many safety-related incidents can be traced to poorly designed and prepared operating instructions and maintenance manuals, as well as inadequate warning labels. When these items are prepared, designers should consider just who will be using their products or machines. Warnings should be permanently affixed in bright colours near all danger points but in locations where they will be readily seen. If possible, these warnings should be permanently attached to the product in an indestructible form. Pictorial warnings have been found to be more effective than written messages.

5. **Design safe tooling and workstations.** One of the most common industrial injuries involves workers inserting a hand into a machine closure point while it is closing. These accidents occur in spite of machine guards because a guard can be circumvented by an enterprising worker, thus making it useless. It is better to have integral or built-in systems that prevent irregular actuation. If correctly designed, such systems can be almost impossible to disable.

6. **Consider maintenance needs.** There are nine basic rules that should be followed to make certain that a product can be safely maintained and repaired.

    (a) *Accessibility.* Parts should be arranged so that the most frequently replaced items can be reached with a minimum of effort.

    (b) *Standardization.* If possible, interchangeable parts, set in common locations, should be used on different models of the product.

    (c) *Modularization.* It may be possible to replace an entire unit without the need to repair malfunctioning individual components. This technique is particularly useful for field repairs.

    (d) *Identification.* Marking, coding, or tagging products makes them easier to distinguish. Colour coding of electrical wiring is a good example.

    (e) *Safety considerations.* When the product is being serviced or repaired, the worker must not be placed in an unsafe situation. For example, sufficient clearance should be provided, particularly in dangerous locations, such as around high voltage sources.

    (f) *Safety controls.* Prevent the machine from being turned on remotely (from another switch location) while maintenance is performed.

(g) *Storage areas.* Provide separate storage areas for special tools, such as those needed for adjustment.

(h) *Grounding.* Ground the machine to minimize electrical hazards.

(i) *Surfaces.* Provide proper walking surfaces to prevent accidents due to slipping or tripping.

7. **Provide clear indications of danger.** Anticipate incorrect assembly, installation, connection, or operation, and prevent them through design. Prevent a malfunction in any single component or sub-assembly from spreading and causing other failures. These techniques are part of fail-safe design.

8. **Control the build-up of energy.** High-energy densities increase the possibility of accidents and should be avoided if possible. For example, include appropriate temperature relief valves or appropriate sensors with automatic shut-off facilities when out-of-range conditions are detected.

9. **Recalls.** The final remedial action is the product recall, as a result of recognizing a real or potential hazard after a device, structure, or system has been put into operation. Action is taken to prevent a potential hazard from becoming a reality. Recalls may be instituted for one or more of five principal reasons:

    (a) Analysis reveals the presence of a potential hazard that can result in a pattern of serious incidents.

    (b) Reports are received from users or others of unsafe conditions, unsafe incidents, or unsafe product characteristics.

    (c) An incident reveals a previously unforeseen product deficiency.

    (d) A government standard or similar regulation has been violated.

    (e) The product does not live up to its advertised claims with regard to safety.

4 Eliminating workplace hazards

People are often required to work in environments where conditions are less than ideal. Such work may expose them to a variety of job-related stresses that can affect their health over a period of time or increase their chances of becoming involved in an accident. The definitions of stress and strain in this context are different from conventional engineering terminology. *Stress* refers to any undesirable condition, circumstance, task, or other factor that impinges on the worker. *Strain* refers to the adverse effects of these stress sources (or stressors) on performance, safety, or health. For example, extremely high (or low) values of temperature, relative humidity, workplace lighting as well as excessive amounts of noise and vibration, are typical stresses that workers may encounter.

Still another important area of concern for both workers and product users relates to chemical contaminants that may be present in production environments or released by products. As a minimum, the designer should be fully aware of any hazards that may be associated with the product. The designer should know how to measure them, recognize the symptoms they produce, and understand the basic steps needed to reduce or eliminate any associated risks to health or safety. Such measures can be applied to either the source, the path, or the user, as explained below.

- *Source control* refers to techniques such as capturing, guarding, enclosing, insulating, or isolating a suspected hazard.
- *Path control* means increasing the distance between the source of the hazard and the receiver. Techniques that are useful here are muffling for noise, grounding for electricity, and improved ventilation to remove toxic byproducts from the air.
- *User control* refers to providing personal protective equipment to workers, thus reducing their exposure to hazards. Also, schedules can be modified to reduce stress, and improved training can be obtained. A general scheme showing examples of control techniques is shown in Table 19.1.

Another common approach used extensively in areas of product safety involves the use of checklists. Frequently such lists take the form of specific questions addressed to the designer. These are intended to prompt him or her to investigate the most common hazards associated with the product's design, choice of materials, manufacturing processes, and functional and maintenance requirements. Examples of checklists are given in Table 19.2 for machine design and Table 19.3 for checking hazards in the workplace.

5 Cost-benefit justification of safety issues

Generally there is a cost justification of safety programs, although the savings from a safety program are often difficult to predict precisely. They may come from a number of different areas:

- Reduced worker's compensation insurance costs.
- Improved productivity from reduced error rates.
- A more stable and content workforce.
- More consistent output, resulting in better quality control.
- Less government involvement to enforce safety standards.
- Confidence in the operating equipment, yielding higher output.
- Improved communication and employee morale.

Table 19.2. Hazard checklist for machine design.

1. Is the machine designed so that it is impossible to gain access to hazard points while the power is on?
2. Are the controls located so that the operator will not be off-balance or too close to the point of operation whenever actuation is required?
3. Are the power transmission and fluid drive mechanisms built as integral parts of the machine so that the operator is not exposed to rotating shafts?
4. Is the machine designed for single-point lubrication?
5. Are mechanical rather than manual devices used for holding, feeding, and ejecting parts?
6. Are there automatic overload devices built into the machine? Are fail-safe interlocks provided so that the machine cannot be started while it is being loaded, unloaded, or worked on?
7. Is there a grounding system for all electrical equipment?
8. Are standard access platforms or ladders provided for the inspection and maintenance of equipment? Are walking surfaces made of non-slip materials?
9. Are equipment components designed for easy and safe removal and replacement during maintenance and repair?
10. Are all corners and edges rounded and bevelled?
11. Are all sources of objectional noise minimized?
12. Are all control knobs and buttons clearly distinguishable and guarded so that they cannot be accidentally activated?

6 Codes and standards

The word *standards* has a general meaning in everyday use, but in the context of engineering design, the term refers to documents describing rules or methods that serve as models of professional practice. Written standards are published by technical societies, as discussed in Chapter 4, and by national organizations but may also be published by governments and commercial organizations. Following a standard in design may be optional but is evidence that the work has been conducted to a professional level of competence.

Codes, on the other hand, are rules or laws, or collections of them, by which a national or local government requires specific practices to be followed, including adherence to particular standards.

National standards organizations include the following:

Table 19.3. Hazard checklist for workplace layout.

1. Design equipment so that it is physically impossible for the worker to do something that would hurt himself, herself, or others.
   Examples of such protective design include:
   (a) A rotary blade that will not start unless a guard is in place.
   (b) The inclusion of interlocks to prevent operation of a machine unless the operator's limbs and body are in a safe position.
   (c) One-way installation: connecting pins that are asymmetrical so that they will fit into a connector in only one way.
2. Cover or guard any moving parts of machinery that could cut a worker or fly off.
3. Make sure that the plant is thoroughly surveyed to detect the presence of noxious gases or other toxic air contaminants.
4. Provide a sufficient number of conveniently located fire extinguishers and an automatic sprinkling system.
5. Use non-slip surfaces on floors and stairways. Eliminate steep ramps used with rolling devices, such as forklifts.
6. Use reliable equipment that will not fail at unscheduled times.
7. Eliminate design features associated with accidents; re-design workspace to eliminate awkward postures, to reduce fatigue, and to keep workers alert while performing repetitive tasks.
8. Label hazards clearly and conspicuously.
9. Provide warning devices.

- Canadian Standards Association (CSA),
- American National Standards Institute (ANSI),
- Deutsches Institut für Normung (DIN),
- Association Française de Normalisation (AFNOR),
- British Standards Institute (BSI).

The International Standards Organization (ISO) is a non-governmental body that creates and approves standards in all subject areas except electrical and electronics, which are handled by the International Electrotechnical Commission (IEC).

6.1 Finding and using safety codes and standards

Many codes and standards have been developed over the years to set minimum acceptable health and safety levels. Using applicable codes and standards for safety and other factors is not only good professional and business practice, but failure to do so may be judged to be professional misconduct, as discussed in Chapter 3. Therefore, you may be faced with determining which codes and standards apply to your work. The following items may be of assistance:

- **Prior practice.** You may be designing modifications to products to which adherence to standards has been documented or to products in the same area. Checking similar prior work or with experienced engineers in the field is a primary source for determining which standards apply. You should check whether standards used previously have since been changed.

- **Search online databases.** The most convenient search method is to search the vast standards databases provided by the CSA and ANSI and more area-specific organizations such as the IEEE and the American Society for Testing and Materials (ASTM). Not surprisingly, given the large number of published standards and applicable codes, there are catalogues listing them, published mainly by the standards organizations such as listed above and by industry-specific bodies such as the Society of Automotive Engineers (SAE) or by commercial testing organizations such as Underwriter's Laboratories (UL). Online catalogues [2, 3] are becoming the standard method for searching for standards. Finally, the cataloguing bodies also provide search services for finding applicable standards.

The statutes and regulations that originate from national, regional and local governments contain the codes covering engineering and other professional work, particularly as it affects public safety. Finding which ones apply to a particular project may involve:

- **Prior practice.** No better source of advice can be found than consulting with an engineer qualified in the specialty and place involved in the project.

- **Search assistance.** Some of the same agencies that list standards also have listings or links to government regulations and statutes. References [3, 4] are good starting places.

The following list is a sampling of codes by government level.

Federal regulations:

    Canada Occupational Safety and Health   Canadian Electrical Code
       Regulations   The National Building Code
    The Labour Code

Provincial regulations:

    provincial building codes   provincial electrical codes
    Construction Safety Act   Drainage Act
    Employment Standards Act   Labour Relations Act
    Fire Regulations   Municipal Act

Planning Act
Industrial Safety Act
Elevators and Lifts Act
Occupational Health and Safety Act
Surveyors Act
Operating Engineers Act
Boilers and Pressure Vessels Act
Environmental Protection Act

Municipal regulations: Each town or city may impose additional regulations or laws or modify codes to suit local conditions.

An engineer beginning work on a new project should ask whether the work may be governed by some code or law. In the case of electrical networks, roads, bridges, buildings, elevators, vehicles, boilers, pressure vessels, and other works affecting the public, the answer will be "yes." Consult with more experienced engineers and determine the codes and laws that apply.

7 Further study

1. In general, all human measurements follow the familiar bell-shaped curve of the Gaussian probability distribution function. There are three basic ways the designer can take into account these differences between people. The designer may

(a) Design for an extreme size, for example, by locating the controls on a machine so that the shortest person can reach them, implying that everyone else can reach them too.

(b) Design for an average; for example, setting the height of a supermarket checkout counter to suit a person of "average" height. Both tall and short cashiers will now have to accommodate themselves to the single height available.

(c) Design for an adjustable range; this is the idea behind the adjustable front seat of an automobile. From its closest to its farthest position, it can accommodate over 99 % of the population.

Is any one of the three design options above always the safest? For each, can you imagine a design situation where it would be safest? Least safe? How would you apply the ideas from Chapters 13 and 14 to achieve method (c) so that a design can accommodate over 99 % of the population?

2. The list below contains various methods of reducing hazards. Can you classify each method as a source, path, or user control method as in Section 4? Do we need additional categories of control?

(a) Prevent the creation of the hazard in the first place. For example, prevent production of dangerous materials, such as nuclear waste.

(b) Reduce the amount of hazard created. For example, reduce the lead content of paint.

(c) Prevent the release of a hazard that already exists. For example, pasteurize milk to prevent the spread of dangerous bacteria.

(d) Modify the rate of spatial distribution of the hazard released at its source, for example, by installing quick-acting shutoff valves to prevent rapid spread of flammable fluid.

(e) Separate, in space or time, the hazard from that which is to be protected. For example, store flammable materials in an isolated location.

(f) Separate the hazard from that which is to be protected by imposing a material barrier. For example, build containment structures for nuclear reactors.

(g) Modify certain relevant basic qualities of the hazard, for example, using breakaway roadside poles.

(h) Make what is to be protected more resistant to damage from the hazard. For example, make structures more fire- and earthquake-resistant.

(i) Counteract damage done by environmental hazards, for example, by rescuing the shipwrecked.

(j) Stabilize, repair, and rehabilitate the object of the damage, for example, by re-building after fires and earthquakes.

8 References

[1] Canadian Centre for Occupational Health and Safety, *Canadian Society of Safety Engineering*, Canadian Society of Safety Engineering, 2001, <http://www.csse.org/> (October 10, 2001).

[2] American National Standards Institute, *NSSN: A National Resource for Global Standards*, American National Standards Institute, Washington, DC, 1998, <http://www.nssn.org/> (November 1, 2001).

[3] Standards Council of Canada, *Database of Canadian, Foreign and International Standards*, Standards Council of Canada, Ottawa, 2001, <http://ww3.scc.ca/> (November 1, 2001).

[4] Department of Justice, *Canadian Legislation*, Department of Justice, Ottawa, 2001, <http://www.legis.ca/> (November 1, 2001).

Chapter 20

Safety, risk, and the engineer

Managing and reducing *risk* and increasing its opposite, *safety*, are paramount engineering responsibilities. Several advanced, computer-oriented techniques are available for analysing hazards and helping to create safe designs. This chapter discusses the following techniques for risk analysis and management:

- checklists,
- operability studies,
- failure mode analysis and its variations,
- fault-tree analysis.

1 Evaluating risk in design

In any industry, it is natural for specifications to evolve to require more demanding specifications, lower cost, or both compared to previous generations of similar designs. Hazards tend to increase because of factors such as greater complexity; greater use of toxic and dangerous substances; higher speeds, temperatures, or pressures at which equipment operates; and more complex control systems. In addition, consumers have an increased awareness of risk and are demanding higher standards of safety. The design engineer is caught between the increased hazards and the higher demands for safety.

Risk factors must be used in evaluating alternative designs. This consideration is important at every stage of the design but is particularly relevant during design reviews, usually held near the end of the project. The design review is a formal evaluation meeting in which engineers and others examine the proposed design and compare it with the design criteria. The review must include a final check that the design has no serious hazards. Unsafe designs do not occur because questions could not be answered; they usually occur because questions about safety were not asked.

The necessity of answering risk-related questions leads to the study of risk management, which is a structured approach for analysing, evaluating, and reducing risk. After identifying and ranking the risks, risk-reduction strategies are applied, starting with the most beneficial and continuing until things are "safe enough," the risk has been reduced to an "acceptable level," or until available resources are exhausted. This process is examined in the following sections.

2 Risk management

A hazard is anything that has the potential to injure, to cause death, to destroy property, to waste financial resources, or to cause any other undesirable consequence. The purpose of risk management is to reduce or eliminate the danger caused by hazards.

We must be able to identify the hazards that are present, estimate the probability of their occurrence, generate alternative courses of action to reduce or eliminate the probability of occurrence, and, finally, we must act to manage the risk. Therefore, risk management can be viewed as a three-step process, involving analysis, evaluation or assessment, and decision-making. The analytical methods in this chapter deal mainly with analysis, but all three steps are described briefly below:

1. **Analysis:** The first step, risk analysis, has two parts: identifying hazards or their undesirable consequences and estimating the probabilities of their occurrence. Hazard identification and probability estimation require logic, deduction, and mathematical concepts, so risk analysis is objective and mathematical. Once the hazards and their occurrence probabilities have been identified, the consequent damage or injury must be evaluated.

2. **Risk evaluation (assessment):** In the second step, alternative courses of action to reduce hazards must be generated, their costs and benefits must be calculated, the risk perceptions of the persons affected must be assessed, and value judgements must be made. Therefore, risk evaluation is less objective than risk analysis. Determining the various options and their acceptability is a tricky and often debatable step in the process. History and experience play a major role. Public perceptions of a product can change: a previously acceptable level of emissions, for example, may become unacceptable. When all these factors are brought together, the evaluation forms the basis for management decisions to control or reduce the risk. It provides decision-makers with possible alternative courses of action and with the necessary information for making informed decisions.

3. **Management decisions:** The final stage of risk management involves selecting the risks that will be managed, implementing these decisions, allocating the required resources, and controlling, monitoring, reviewing, and revising these activities.

In design, the term *good practice* means following established methods for safe design or construction. Good practice is often embodied in industry standards and government regulations that may be mandatory. However, whether mandatory or not, adherence to good practice does not guarantee that a product will be safe in the hands of the ultimate user or that a plant to carry out a process can be designed, built, operated, and decommissioned with an acceptable level of risk. From the perspective of liability, adherence to good practice is the minimum acceptable level of care expected, and more advanced risk-management methods are usually essential.

There are many methods that can be used to identify hazards and to evaluate risk; they range from the informal to the highly structured. Some methods merely identify hazards; others help to determine the consequences or the probabilities of the consequences occurring, or both. Still other techniques extend into risk evaluation because they rank the risks according to certain criteria. The techniques are not mutually exclusive, and many companies modify these standard techniques for their own purposes. In some cases, a combination of methods may be best. Four frequently used methods for risk analysis are explained briefly in the next section.

3 Analytical methods

The following four analytical approaches are commonly used in risk analysis, which is sometimes also called *safety engineering:*

- checklists,
- hazard and operability studies,
- failure mode analysis,
- fault trees.

Each of these methods will be briefly explained and discussed.

3.1 Checklists

Checklists are particularly useful when a design has evolved from a previous design for which all hazard sources were carefully listed, so that the consequences of the evolution are easy to identify. However, checklists must be used with discretion. Preparing a comprehensive, useful checklist can be much work, and if the product or process changes, the checklist must be revised. Hazards that are not specifically mentioned in the checklist will be overlooked, and hazard definitions can change if new information becomes available or if perceptions of risk change.

Checklists should be developed as early in the project as possible. In fact, ideally they should be part of the design criteria. The checklists in Table 19.2 and Table 19.3 are examples of a typical format. These examples as well as the generic headings suggested below can be used as a basis for developing a design-review checklist for a specific product or process.

Typical headings for a safety checklist for design review:

1. General standards and standardization
2. Human factors: displays and controls
3. Maintainability in service

4. Instruction for users and maintenance personnel
5. Packaging, identification, and marking
6. Structural materials
7. Fittings and fasteners
8. Corrosion prevention
9. Hazard detection and warning signals
10. Electrical, hydraulic, pneumatic, and pressure sub-systems
11. Fuel and power sources.

Checklists are built on past experience. For new or radically altered products and processes, checklists should not be used as the sole risk-management tool, since methods that identify hidden or inherent hazards and their consequences are required. These methods are called *predictive techniques* and are important in today's litigious climate. Innovation usually increases risk, and innovation would be stifled if such techniques were not available. Three commonly used predictive techniques follow.

3.2 Hazard and operability (HAZOP) studies

Methods similar to the description in this sub-section are used mainly in the process industries to uncover hazards and problems that could arise during plant operation. The HAZOP technique could be called *structured brainstorming*. The technique is applied to precisely specified equipment, processes, or systems; therefore, the design must be beyond the concept stage and in a more concrete form before the technique can be used. Variations of the HAZOP technique can be used at all stages of the life of a plant, such as at preliminary design, final design, start-up, operation, and decommissioning.

The study is carried out by a team of experts who, together, have a full understanding of the process to be examined, such as the basic chemistry, process equipment, control systems, and operational and maintenance procedures. To carry out the study, complete information is needed: flowcharts, process diagrams, equipment drawings, plant layouts, control system descriptions, maintenance and operating manuals, and other information as necessary.

Some definitions and concepts used in HAZOP studies include the following:

- *nodes:* the points in the process that are to be studied
- *parameters:* the characteristics of the process at a node
- *design intent:* how an element in the process is supposed to perform, or the intended values of the parameters at each node
- *deviations:* the manner in which the element or the parameters deviate from the design intent
- *causes:* the reasons why the deviations occur

- *consequences:* what happens as a result of the deviations
- *hazards:* the serious consequences (note the different meaning of hazard in this context)
- *guide words:* simple words applied to the elements or parameters to stimulate creative thinking to reveal all of the possible deviations

In HAZOP analysis, the first step is to identify the nodes in a plant process; each node is then studied, as in the following.

Example 1. A heat exchanger.

Consider a process that includes a heat exchanger in a duct carrying hot gas. Water flows through the coil in order to cool the gas. The water flow into the heat exchanger would be one of the study nodes. The parameters at this node would include flow, temperature, and pressure. The design intent at this node might be to provide water at a flow rate of, say, $0.1 \, \text{m}^3/\text{s}$ at a temperature between $10\,°C$ and $25\,°C$ and pressure not to exceed 10 atm. A set of guide words would be applied successively to the parameters. To illustrate, consider the guide words "NO FLOW." The cause of "NO FLOW" could be a closed valve, a broken pipe, a clogged filter, or the failure of the municipal water supply. Each of these causes is realistic, so the deviation could occur and the consequences of the deviation must be examined. The possible consequences could include vaporising the water in the heat exchanger coil, thus producing dangerously high pressure or allowing the temperature of the gases to rise, yielding dangerously high rates of reaction further downstream in the process. Both of these consequences are serious and warrant attention.

The other parameters at the node—temperature and pressure—would also be studied. All other nodes in the process would be studied in a similar manner.

The prime purpose of HAZOP studies is to identify hazards and operational problems. The hazards and problems must be well documented for subsequent attention, and there must be an effective follow-up to ensure that the HAZOP decisions are implemented.

3.3 Failure modes and effects analysis (FMEA)

Techniques similar to what is described here were developed by reliability engineers to determine the reliability of complex systems. With the addition of the assessment of criticality, a technique called "failure modes, effects, and criticality analysis" (FMECA) results. The use of this technique has spread to many disciplines. The technique is inductive, and the analysis is carried out by completing a specifically defined table following logical rules [1]. Full details of the component, equipment, process, or complex facility to be analysed must be available, including drawings, design sketches, standards, product specifications, test results, or other information.

The FMEA method is a "bottom-up" process, that is, it traces the consequences of low-level failures to higher levels. The definition of the bottom level depends on the desired resolution of the analysis. For instance, for a chemical plant with several independent processes, the bottom might be an individual process in a reactor. If a single process were considered, the bottom might be the individual pieces of equipment supporting that process. If a specific piece of equipment were to be analysed, say a pump, the bottom might be the component parts: the housing, seals, impeller, coupling, motor, and motor control.

FMEA process: The steps followed in an FMEA or FMECA analysis on a piece of equipment are as follows:

1. List each of the components or sub-systems in the unit.
2. Identify each component by part name and number.
3. Describe each component and its function.
4. List all of the ways (modes) in which each component can fail.
5. Determine for each mode of failure the effect of that failure on other components and on the unit as a whole. Enter the information in the table.
6. Describe how the failure mode can be detected.

FMECA process: In addition to the FMEA analysis, criticality analysis continues with the following:

7. Indicate the action to be taken to eliminate the hazards and identify who is responsible for taking that action.
8. Assess the criticality of each failure mode and its probability of occurrence.

Many forms of tables have been developed for use with FMEA and FMECA techniques. The measure of criticality is the seriousness of the consequence, which may range from a simple requirement for maintenance, through property loss or personal injury, to catastrophe. In the FMECA method, a measure of the consequence is usually combined with the estimated probability of occurrence to weigh the risk associated with each failure.

The potential difficulty in using bottom-up methods for complex products or systems is the large effort required by the large number of components. Frequently, another technique, such as the qualitative fault-tree analysis described in Section 3.4, can be used to narrow the scope of the analysis. The FMECA method is effective in determining the consequences of single component failures; it is not suitable for determining the effect on the system of the concurrent failure of several system elements. The results of the method are qualitative, or semi-quantitative if the failure modes are ranked according to the criticality of their consequences. The analyst must be familiar with the components of the system under study and must be able to determine all of the possible modes of failure and their consequences.

3.4 Fault-tree analysis

The method described here was developed by Bell Laboratories for the U.S. Air Force, and the technique has since been refined and used for many different purposes. In contrast to FMEA, fault-tree analysis is a "top-down" process. The starting point in a fault-tree analysis is a single event, called the top event. The first step in the analysis is to determine all events that could cause the top event. These contributor events are then analysed, in the same manner, to determine the events upon which they are dependent. The process is continued until events are reached that do not depend on other events or need not be further broken down. The events that are not dependent on other events are called *primary*, or *basic fault events*. The events that are not further broken down are called *undeveloped events*. Events can be human errors or failures of equipment. Fault-tree analysis can be used at any stage in the evolution of a product or process. The events and their relationships are usually represented diagrammatically using standard logic symbols. This results in a tree-like diagram [2], which gives the method its name.

Example 2. A hair dryer.

The common hand-held hair dryer can be used to illustrate the fault-tree technique. Assume that the dryer is made from non-conducting plastic, has no metal parts that extend from the inside of the dryer to the outside, and has a two-conductor power cord. This is typical construction for hair dryers in the consumer market. Assume also that the dryer is not connected to a power supply protected by a ground-fault interrupter (GFI) circuit breaker. Although GFI circuit breakers are normally required in bathrooms, it cannot be assumed that the dryer will always be plugged into a GFI circuit.

In this example, the top event is the electrocution of the user. The purpose of the analysis is to determine how this top event could occur, and then to take steps to prevent it from happening, or if total prevention is not possible, to reduce the likelihood of such a disaster occurring.

The fault tree is shown in Figure 20.1. The top event will occur only if all of the inputs to the AND gate are present; that is, all of the events in the row immediately below the AND gate occur. The shape of the leftmost box identifies an event that is expected to occur. The dryer uses the normal 120 V house supply, a dangerously high voltage. Two events in the first row are enclosed in diamonds and not developed further. The two remaining events in this row depend on outputs from the OR gates. An OR-gate output occurs if any of its input events occur. The inputs to the OR gates are basic fault events and are enclosed in ellipses.

252 Chapter 20 Safety, risk, and the engineer

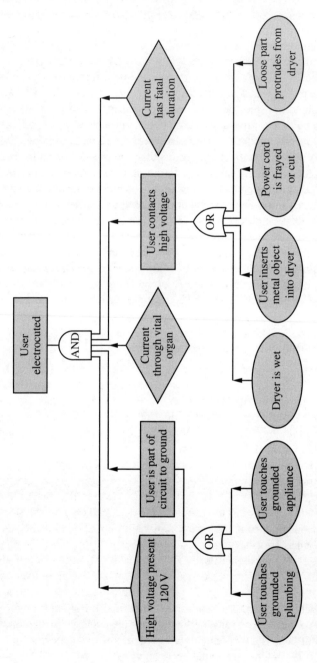

Fig. 20.1. Partial fault tree for the hair dryer in Example 2. The two diamonds represent undeveloped sub-trees.

The logic tree presents the results of the analysis. Let us look at some qualitative conclusions that can be reached. If the top event is undesirable, then the presence of AND gates is favourable, since all of the inputs to the AND gate must be present for the higher event to occur. In a simple tree such as Figure 20.1, the basic or undeveloped events that must occur for the top event to happen can be determined by inspection. Guided by the information in the fault tree, steps can be taken to make the product safer for the user.

Some of the basic fault events depend solely on the actions of the user and cannot be prevented by design changes. However, instructions for the proper use of the dryer and conspicuously placed warnings about the hazards of misuse, as discussed in the previous chapter, should decrease the probability of injury and lessen the potential liability of the engineer. In the previous example, inserting fingers or metal objects through the dryer openings can bring a person into contact with high-voltage components; this event can be made more difficult by placing the high-voltage components farther away from the openings. The insertion of objects through openings in electrical appliances is a well-recognized hazard, and Underwriter's Laboratories [3] and others have standard probes to be used in the certification of products. Of course, this is only one of the requirements for certification.

The analysis described above is qualitative in nature; however, the next step in the analysis is usually quantitative. If failure probabilities for the basic fault events are known, the probability of the top event occurring can be calculated [4]. The failure probabilities for many standard components are available from reliability handbooks and other sources. Tests can be conducted to determine failure probabilities for components or systems for which data is not available. In a more complicated tree, Boolean algebra must be used to determine the sets of basic events that must all occur for the top event to happen. Computer programs are available for the construction and analysis of fault trees.

4 Safety in large systems

Systems containing many interconnected parts present engineering problems of a different type than basic products, artifacts, or static structures. Systems such as those used for space navigation, vehicle control, aircraft control, industrial processes such as chemical or petroleum refining or power generation, indeed, almost any system with a computer embedded in it will have distinguishing features that render safety analyses difficult.

The first such distinguishing feature is complexity. A system can be thought of as a graph, that is, a set of nodes representing system variables or parameters, joined by branches representing processes or basic sub-systems. Then if the number of branches is of the same order as the number of nodes, the system is said to have low complexity, whereas high complexity corresponds to cases where the number of branches is large compared to the number of nodes.

Another factor of importance is the coupling between parts of the system. In a tightly coupled system, a perturbation in one part may cause changes in many other parts, while in a loosely coupled system perturbations in one part have little effect on other parts. Coupling is related to complexity, since complex systems have the potential to be more tightly coupled.

From a safety standpoint, the system response rate is very important, particularly as it affects the opportunity for operator intervention. Fast-response runaway processes require very careful safety measures, while a simple alarm may suffice for a process with very slow response time.

The concepts of feedback and stability are related to response time and complexity. Interdependence, through information and control feedback paths, can lead to unstable modes of operation. The avoidance of these modes is a prime consideration in safety design.

Finally, a design is considered robust if it brings about fail-safe or fail-soft conditions in the event of trouble. A fail-safe system can suffer complete loss of function without any attending damage. A fail-soft system may suffer loss of functionality in the event of failure but retains a minimum level of performance and safety, such as the possibility of manual control.

In conclusion, a large system may be reasonably safe if it is not too complex and is loosely coupled, slow to respond, stable, and robust to part failures. Safe operation will be difficult to achieve for a complex system that is tightly coupled, with fast response rate, many feedback paths, and sensitivity to part failures. Since the design and application of increasingly complex computer-based and communication-based systems is the responsibility of the professional engineer, a control system analysis course is found in almost every engineering discipline. PEO provides recommendations in a telecommunication guideline and a software guideline to assist in the design, implementation, testing, and documentation of these safety-critical components.

5 System risk

A proper engineering management of risk requires the quantization of the basic variables involved. The risk of a hazardous event can be defined as a function of the probability of the event and the cost of an occurrence of the event. Because high-cost, high-probability events tend to be designed out of systems, higher probability will typically occur with the lower costs, and lower probability with the higher costs. Indeed, the area of the most intense debates is where the probability is very small but the cost is very severe, particularly when the cost applies to a very few and is not shared by a large number of beneficiaries of the process. Consider, for example, safety improvements on highways with low traffic flows: accidents may have a low probability but may involve fatalities when they occur.

One definition of risk r is the product of the probability p of an event and the event cost c. Thus,

$$pc = r, \tag{20.1}$$

or, taking the logarithm of both sides,

$$\log(p) + \log(c) = \log(r), \tag{20.2}$$

so that the constant-r loci are straight lines when the above function is plotted using log-log scales, as illustrated in Figure 20.2. The figure gives a simple interpretation of the trade-off between event cost and probability; the difficulty in using such definitions is in obtaining precise estimates of these quantities.

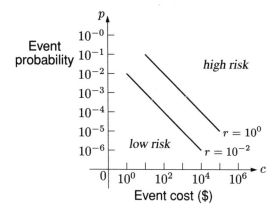

Fig. 20.2. Illustrating lines of constant risk. Events in the upper right part of the chart have high risk; low-risk events are in the lower left.

The probability of a hazardous event must be determined and defined relative to the population exposed to it. Some hazards have very low probabilities when averaged over the total population but may be more probable if only the participating population is considered. For example, the mortality of astronauts and coal miners must be evaluated relative to the number of participants, not the population as a whole. Similarly, when accident rates are compared for different modes of travel, they are often expressed in accidents per passenger-mile to give a defensible comparison.

The cost of event consequences may be even harder to define. Some costs, such as the cost of warranty replacement or repair and the cost of maintenance and service, may be relatively easy to estimate. However, the costs borne by others, including the total cost to society, may be impossible to estimate.

6 Expressing the costs of a hazard

In product safety analyses, costs of consequences are normally measured in monetary terms. The cost includes the legal costs associated with responsibility for injury. Legal costs can be very high and are an important component affecting the high malpractice insurance rates in health care and for determining liability insurance costs in engineering and other professions.

On a societal level, the total costs of the consequences of an event are not as simple to predict. Resource depletion is frequently overlooked; risk evaluations rarely include resources held in common with other members of society, such as fish stocks, forest areas, fresh water, clean air, and non-renewable resources that are not privately owned. Moreover, since it is not common practice to take these assets into account when evaluating a national economy, apparent economic progress can be associated with the impoverishment of natural assets.

The risk of a particular hazard to the human population is usually termed *mortality:* the probability of death per year for exposed members of the population, or as loss of life expectancy expressed in days. The latter has the advantage that safety measures such as smoke alarms and air bags can be given a negative loss figure; that is, the gain in life expectancy by their use can be evaluated.

The wealth of an industrialized society depends on the creativity of its engineering sector, and technical advances almost always involve some risk. For example, in the last 160 years, life expectancy at birth in Canada has increased by a factor of 2. This gain results primarily from improvements in safety and sanitation, communication, transportation, water supply, food processing, and from medical and health improvements. The loss in life expectancy from the hazards of industrialization are small relative to the gains. Nevertheless, it is human nature, and thus engineering nature, to seek improvement. Large-scale projects now must be given a life-cycle cost analysis so that the designers see the full cost of the plant, including decommissioning and disposal of wastes, and not merely the construction and start-up costs.

7 Further study

1. Suppose that you have designed a portable compressed-air supply system, as shown in Figure 20.3, for filling underwater scuba tanks and also for driving pneumatic tools in garages and factories. The system consists of an electric motor, an air compressor, an air tank, a regulator, and a pressure relief safety valve. The regulator contains a pressure sensor connected to a power switch. The motor is switched on when the tank pressure drops below a fixed value and off at a slightly higher pressure. The pressure-relief valve is set to open at a higher pressure than the regulator motor-off pressure. Explosion of the air tank is clearly a dangerous hazard, since it could cause injury or death. Three events are estimated to possibly lead to a tank explosion:

- an internal tank defect such as poor welding,
- an external cause, such as a plant vehicle colliding with the tank, and

Section 7 Further study 257

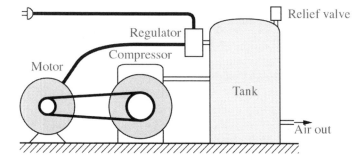

Fig. 20.3. Compressed-air supply unit.

- excess pressure in the tank. The excess pressure can be caused only if both the control unit and the pressure-relief valve fail, as described below.
  - The pressure sensor might fail to shut off the motor because of
    · switch failure, or
    · human error, such as the switch set incorrectly or propped open deliberately.
  - The pressure-relief valve might fail to open because of
    · mechanical valve failure, or
    · human error, such as setting the valve to incorrect pressure or locking it shut.

Construct the fault tree for the system described above, for the case in which explosion of the air tank is the top event.

Answers to problems:

1.

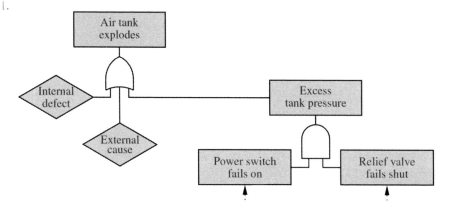

Fig. 20.4. Part of the fault tree for Problem 1.

8 References

[1] B. Dodson and D. Nolan, *Reliability Engineering Handbook*, Marcel Dekker, Inc., New York, 1999.

[2] H. Kumamoto and E. J. Henley, *Probabilistic Risk Assessment and Management for Engineers and Scientists*, IEEE Press, New York, 1996.

[3] Underwriter's Laboratories, *Household Electric Personal Grooming Appliances, Safety Standard 1727*, Underwriter's Laboratories, Northbrook, IL, 2000.

[4] J. R. Thomson, *Engineering Safety Assessment: An Introduction*, John Wiley & Sons, Inc., New York, 1987.

Index

abstract, *see* reports, abstract
academic requirements, 20, 21
accountability, 6, 233
accreditation, 20, 21, 25, 44, 45, 50, 183
accuracy, 125, 128
adjective, 75, 76
admission requirements, 20, 22, 23, 25
   documentation, 23
   examinations, 23
   experience, 22, 23
   time limits, 23
adverb, 75, 76
alcohol, 54
American Gear Manufacturers Association (AGMA), 47
American Institute of Chemical Engineers, 47
American Institute of Electrical Engineers, 43
American Institute of Mining, Metallurgical and Petroleum Engineers (AIME), 47
American National Standards Institute (ANSI), 234
American Society for Engineering Education (ASEE), 47
American Society of Civil Engineers (ASCE), 47
American Society of Mechanical Engineers (ASME), 47

analysis, 199
analytical ability, 10
Asia, 184
Asimow, M., 187
assignments, hints, 53
Association of Consulting Engineers of Canada, 47
Association of Professional Engineers of Ontario (APEO), 17
Association of Professional Engineers of Ontario (PEO), 17
Association, provincial, 4, 5, 13, 15, 20, 21, 23–25, 41, 42, 44, 45
Australia, 21

Bar Association, 24
behaviour
   academic, 30
   ethical, 86
   human, 186
   professional, 30
   threatening, 30
   unethical, 28
   unprofessional, 27, 31
bias, 125, 127, 128
bibliography, 92
bids, 64
brainstorming, *see* creativity, brainstorming
bribe, 27
Britain, 17, 42
by-laws, 17–19, 24, 33

calculations, 107
    format, 107
Canada, 4, 5, 7, 12, 14–17, 20–22, 35, 42, 43, 45, 116, 126, 184, 208, 211, 212, 214–217, 256
Canadian Council of Professional Engineers (CCPE), 44
Canadian Academy of Engineering, 47
Canadian Alliance Against Software Theft (CAAST), 35
Canadian Council of Professional Engineers (CCPE), 4, 17
Canadian Engineering Accreditation Board (CEAB), 21, 44, 45, 183
Canadian Engineering Qualifications Board (CEQB), 22, 44
Canadian Federation of Engineering Students (CFES), 47
Canadian Geotechnical Society (CGS), 46
Canadian Institute of Mining and Metallurgy (CIM), 46
Canadian Institute of Surveying, 43
Canadian Intellectual Property Office (CIPO), 204, 207, 208, 212
Canadian Medical and Biological Engineering Society, 46
Canadian Nuclear Society, 46
Canadian Society for Chemical Engineering (CSChE), 46
Canadian Society for Civil Engineering (CSCE), 46
Canadian Society for Engineering Management, 46
Canadian Society for Mechanical Engineering (CSME), 46
Canadian Society of Agricultural Engineering (CSAE), 46
Canadian Society of Professional Engineers (CSPE), 47
CANCOPY, 213

carpenter, 6
central limit theorem, 167
certificates of authorization, 23
Certified Engineering Technician, 6
Certified Engineering Technologist, 6
chart, 58, 76, 88, 99, 102, 201
    bar, 102, 221
    flow, 58, 102, 248
    Gantt, 182, 221, 222, 230, 232
    geographical, 103
    organizational, 102
    pie, 102
    risk, 255
Chartered Engineer, 42
citation, 89–92, 97
clarity, 69
Code of Ethics, 2, 18, 19, 27, 28, 30–32, 38, 39
codes and standards, 240
    federal, 242
    municipal, 243
    provincial, 242
    safety, 241
coffee, 52, 54
College of Physicians and Surgeons, 24
comma, 73, 74, 76
    splice, 76
community college, 5, 6
computer
    circuits, 9
    errors, 34
    expertise, 10
    hardware, 7, 9
    lab, 2
    program, 27, 34, 100, 186, 210, 213, 230
    revolution, 4, 14
    science, 7, 9
    software, 9, 34, 35
    subjects, 7
    systems, 7, 36
    viruses, 34, 36
confidence level, 151, 165, 171, 172

conjunction, 75
constraints, 186
consulting engineer, 4, 39, 42
copying, *see* plagiarism
    fair dealing, 213
    illegal, 203, 213
copying, unauthorized, *see* software, piracy
copyright, 35, 59, 86, 109, 182, 204, 206, 207, 210, 212–214, 218, 219
    Act, 212, 213
    CANCOPY, 213
    computer programs, 213
    convention, 212
    fair dealing, 213
    foreign, 212
    infringement, 212
    registration, 212
    reverse engineering, 213
Corporation of the Seven Wardens, 37
correlation, 153, 165, 176, 177, 179
counselling, 49, 55
course outline, 51, 54
cover letter, *see* reports, transmittal letter
CPM, 221, 222, 225, 227, 228, 230
    critical path, 227
    event times, 227
    summary, 230
creativity, 182, 195–197
    blocks, 198
    brainstorming, 195, 197, 198, 248
    brainwriting, 198

deciles, 161
decision-making, 11, 153, 182, 187, 192, 195, 199–201, 246
    computational, 200
degree programs, 6, 7, 9, 10, 20, 21, 44, 45, 66
design
    checklists, 247

communication, 188
criteria, 186
cycle, 187
definition, 183
evaluation, 188
examples, 188
feasibility, 185
guidelines, 234
human interface, 186
importance, 183
industrial, *see* industrial design
need, 185
optimization, 183, 188, 192
preliminary, 187, 191, 248
problem definition, 185
process, 182–184
responsibility, 233
review, 188, 245, 247
risk, safety, 182, 233, 245
specifications, *see* specifications
synthesis, 187, 195
team, *see* teams, design
diagrams, *see* reports, diagrams
dictionary, 24, 69–71, 76, 199
digits, significant, 129, 130
dimensions, 116
    fundamental, 116
disasters, 15, 51, 233
disciplinary action, 27, 32, 35–37, 107, 234
    code, 30
    committee, 32
discipline, *see* engineering, disciplines
discrimination, 31
disputes, 32
    email, 63
distribution, 152
    sample, *see* sample, distribution
    skewed, 157–159, 161, 169, 179
documents
    back matter, 90
    body, 87
    front matter, 82

technical, 58, 59
types, 59
drawings, 36, 58, 70, 93, 100, 109, 188, 191, 211, 212, 214, 248, 250
drugs, 31, 54

electrician, 6
electronic mail, *see* email
email, 62
    disputes, 63
    privacy, 63
    security, 63
embezzlement, 16
employment contracts, 207
engineer
    competent, 10
    definition, 3–5
    role, 5
Engineer in Training, 22
engineering
    admission, *see* admission requirements
    aerospace, 10
    agricultural, 10
    and mathematics, 10
    biochemical, 10
    biosystems, 10
    building, 10
    calculations, 58, 99
    ceramic, 10
    challenges, 3, 12
    chemical, 6, 7
    civil, 3, 6, 7, 9
    communications, 58
    computer, 7
    definition, 19
    design, *see* design
    disciplines, 6
    electrical, 6–8, 42
    environmental, 9
    forest, 10
    geological, 9
    industrial, 9
    introduction to, 2, 3
    materials, 10
    measurements, 114
    mechanical, 6, 7, 9, 42
    mining, 10
    petroleum, 10
    physics, 10
    practice, 182
    principles, 19
    profession, *see* profession
    regulation, 16, 17
    safety, 247
    science, 9
    societies, *see* engineering societies
    software, 7, 9
    systems design, 9
Engineering Institute of Canada (EIC), 43, 46
engineering societies, 41–43, 46–48
    advocacy, 47
    Canadian, 46
    choosing, 45
    directory, 46
    history, 41–43
    honorary, 45, 47
    importance, 41, 43
    international, 47
    purpose, 2, 41, 42, 48
    student chapters, 46
Engineering Society of the University of Toronto, 43
Engineers Joint Council, 46
England, 4, 184, 206, 218
errors
    addition, 132
    arithmetic mean, 133
    bias, 127
    computer, 34
    derived, 114, 132, 135, 137, 142–144, 146
    division, 133
    instrument, 127
    law of, *see* law of errors

Index 263

linear estimate, 137
measurement, *see* measurement, error
multiplication, 133
natural, 127
personal, 127
random, 126, 127
range, 141
rounding, 155
subtraction, 132
systematic, 126, 127
ethics
   code, 2, 13, 15, 17–19, 24, 27, 28, 30–33, 36–39, 42
   dilemma, 31, 38
   examination, 23
   problems, 31
   professional, 2, 27, 28, 38, 203
   workplace, 27, 30
examinations
   preparation, 49, 51, 53
   professional admission, 21, 23
   professional practice, 38
   writing, 54
experience
   Canadian, 22
   military, 23
   requirements, 15, 18, 20, 22, 24

fail-safe, 238, 240, 254
fail-soft, 254
failure modes and effects analysis, *see* FMEA
fault-tree analysis, *see* safety, fault tree
figures, *see* graphics
Florman, S. C., 4, 16
FMEA, 245, 249, 250
FMECA, 249, 250
foot, 120
Formula SAE, 48
France, 21

Génie unifié, 9

Gaussian
   distribution, 114, 152, 165–171, 173, 174, 179
   law of errors, 165, 168, 173
global warming, 12, 185
good character, 20
good faith, 29, 32
graphics, 58, 99, 100
graphs
   log-linear, 105, 106
   log-log, 106, 107, 255
   logarithmic scales, 104
   power function, 106
   standard format, 100
   straight-line, 99, 103, 105, 106
Greek alphabet, 77

harassment, 33, 34, 192
hazards, 236, 238, 255
   HAZOP studies, 248
histogram, 162, 173
   bin size, 163, 174, 179
Hong Kong, 21

IEEE Canada, 46
incompetence, 16, 32
industrial design, 204, 214
   Act, 214
   application, 214
Industrial Revolution, 4, 45, 184
Industry Canada, 204
ing., 5
ingenium, 3
Institute for National Measurement Standards, 126
Institute of Civil Engineers, 42
Institute of Electrical and Electronic Engineers (IEEE), 43, 47
Institute of Radio Engineers, 43
Institution of Mechanical Engineers, 42
insurance
   liability, 6, 23, 256
   malpractice, 256

integrated circuit topographies, 204, 216
  Act, 216
intellectual property, 36, 59, 109, 182, 203, 204, 206, 207, 209, 213, 216–218
  employee rights, 207
  employer rights, 207
  ideas, 203
  proprietary, 204
  public domain, 206
interjection, 75
internet, 10, 14, 17, 34–36, 48, 50, 52, 56, 63
Ireland, 21
Iron and Steel Society (ISS), 47
iron ring, 2, 3, 37

jargon, 70, 71
job offer, 31
job placement, 31
job sites, 30
Junior Engineer, 22
justice, natural, 32

key words, *see* reports, key words
Kipling, R., 37

laboratory report, *see* reports, laboratory
Landauer, T. K., 186
law of errors, 114, 168
Law Society, 24
lawyer, 16, 58, 70, 204
lectures, 50–54
lettering, 109, 110
letters, 60
  business, 60
liability, 23, 234, 246, 253
  for computer errors, 34, 35
  lawsuit, 237
licence, 2, 15, 17, 19, 20, 22, 23, 29, 32, 34, 39, 41
  practising without, 32

Lippman, 168
machinist, 6
MacKenzie, H., 16
malpractice, 16, 27, 256
manufacturing, 7, 9, 10, 42, 48, 58, 71, 115, 153, 159, 184, 186–188, 190, 191, 193, 208, 233, 239
Marine Technology Society, 46
mean, 156, 172, 173
measurement, 114, 115
  calibration, 125
  definition, 125
  digital display, 129
  dimensions, 116
  error, 114, 125, 126, 129, 132, 133, 135, 138, 140, 144, 151, 165, 168
  precision, 129, 134
  presentation, 130, 161
  procedures, 135
  representation, 115
  spread, 153
  systems, 116
  techniques, 115
  traceable, 125
  true value, 152
  uncertainty, *see* uncertainty, measurement
  units, 115, 121
median, 156
Member in Training, 22
memos, 60
  format, 60
Minerals, Metals & Materials Society (TMS), 47
misconduct, professional, 27, 32, 33, 35, 37, 39, 233, 242
mode, 158
modifiers, dangling, 76
Moore, G., 104
  Moore's law, 104, 111

National Research Council, 126

National Society for Professional
    Engineers (NSPE), 48
negligence, 33, 37
New Zealand, 21
newton, 115, 120
Newton's second law, 116, 118, 120,
    123, 206
notation
    engineering, 130, 131
    fixed, 129–131
    inexact quantities, 129
    rounding, 131, 132
    scientific, 129–131
    uncertainty, 129, 130
note-taking, 52
noun, 75

oath, engineering, 37
Obligation of the Engineer, 37, 38
Ontario Association of Certified
    Engineering Technicians
    and Technologists, 6
Ontario Society of Professional
    Engineers, 24, 45, 47, 48
open-mindedness, 11
Ordre des ingénieurs du Québec, 4, 17
organizations, engineering, 24
outliers, 177–179

P.Eng., 4, 5, 19
partial derivative, 139, 145, 149
parts of speech, 69, 75
patent, 204, 208, 209
    Act, 208
    application, 210
    Cooperation Treaty, 211
    criteria, 208
    first-to-file rule, 210
    first-to-invent rule, 210
    foreign, 211
    Patent Office Record, 211
    pending, 211
pattern maker, 6
penalty
    academic, 36
    criminal, 212
PEO, 17–19, 22, 24, 27, 28, 30, 32,
    33, 35–38, 43, 234, 254
    Gazette, 32, 37
    software guideline, 254
    telecommunication guideline,
        254
percentiles, 161
person, 73
Petroski, H., 4
physicians, 16
plagiarism, 30, 34, 36, 53, 56, 89, 212
plumber, 6
pollution, 12, 190
pound, 118, 120
precision, 125, 128
preposition, 75
problems
    academic, 55
    personal, 55
    solving, 182, 184, 195, 198, 199
profession, 2, 4, 5, 13, 15–21, 24,
    28–32, 34, 43–45, 184
    self-regulating, *see*
        self-regulation
Professional Engineer, 4, 5, 13, 15,
    17, 19, 39
Professional Engineers Act, 16–19,
    23
Professional Engineers Ontario, *see*
    PEO
professional practice examination, 23,
    28
project
    milestones, 223
    planning, 182, 221–223, 225,
        227, 228, 230
    scheduling, 182, 221, 225
pronoun, 75
property, intellectual, *see* intellectual
    property
proposals, 64

provincial Association, *see* Association, provincial
punctuation, 58, 69, 71, 73, 78, 79, 95

Québec bridge, 15
quality control, 153
quartiles, 161
quotations, 52, 74, 89

range, 159
references, 74, 91, 92, 97, 98, 100
regression, 175, 176
regulations, 18, 20, 32
reports, 64
    abstract, 86
    acknowledgements, 86
    appendices, 92
    back matter, 65, 95
    bibliography, *see* bibliography
    body, 65
    checklist, 95
    citations, *see* citation
    components, 67, 82
    conclusions, 65, 90
    copyright, 212
    cover, 82
    diagrams, 88
    feasibility, 66
    formal, 58, 81
    front matter, 65, 95
    graphics, *see* graphics
    internal sections, 88
    introduction, 65, 87
    key words, 86
    laboratory, 65–68, 218
    lists of figures, tables, symbols, 86
    organizing, 94
    outline, 94
    planning, 94
    preface, 85
    progress, 68
    purpose, 66
    reader, 94
    recommendations, 90
    references, *see* references
    revising, 95
    submitting, 95
    summary, 86
    supporting sections, 95
    table of contents, 86
    technical, 65, 73, 81, 86
    test, 66
    title page, 83
    topics, 58
    transmittal letter, 85
    writing steps, 72, 93
risk, 182, 245, 254
    analysis, 246, 247, 253
    decisions, 246
    evaluation, 245–247
    good practice, 246
    management, 245–248
    mortality, 255, 256
    system, 254
rounding, *see* notation, rounding

safeguarding of life, 19, 33
safety, 182, 245
    analyses, 255
    checklists, 245, 247
    codes and standards, 42, 233
    cost-benefit, 239
    design, 182, 233, 234, 237–239, 245, 247, 253, 254
    environmental, 184
    fault tree, 245, 247, 251–253, 257
    highway, 190, 254
    operability studies, 245
    product, 66
    public, 5, 33, 165, 234, 242
    response rate, 254
    responsibilities, 233
    signs, 234, 236
    travel, 201, 202
    worker, 182, 234
sample, 152

distribution, 152, 155, 169
    mean, 156, 160, 171–174, 178
    standard deviation, 160, 171, 173, 174, 178–180
    variance, 160, 178
scientist, 2, 5, 19, 166
seal, engineer's, 2, 27, 33, 36, 37
self-regulation, 5, 16–19, 24, 32
sensitivities, 139
    relative, 140
sentence
    fragments, 76
    run-on, 76
    structure, 73, 75, 90
    subject, 76
    verb, 76
SI, *see* unit systems, SI
SI prefixes, 129
sickness, 55
significant digits, *see* digits, significant
sketches, 109
skilled worker, 5, 6
skills, 10
    communication, 11, 58, 69
    human, 4
    inventiveness, 11
    key, 69
    management, 9
    mathematical, 10
    problem-solving, 10
    study, 49, 56
    study checklist, 51
    writing, 93
slug, 120
Society for Mining, Metallurgy, and Exploration (SME), 47
Society of Automotive Engineers (SAE), 47, 242
Society of Petroleum Engineers (SPE), 47
software, *see* computer, software
    faulty, 35
    piracy, 34–36, 212, 213

South Africa, 21
specification documents, 63
specifications, 36, 38, 63, 64, 129, 188, 191, 245, 250
spelling, 71, 95, 97, 120
standard deviation, 160
statistics, 114, 151
    applications, 152, 153
    central value, 153, 155
    data presentation, 161
    deciles, 161
    definitions, 151, 152
    descriptive, 153
    error, 152
    inferential, 153
    mean, 156
    median, 156
    mode, 158
    observation, 151
    parent distribution, 152
    parent population, 151
    percentiles, 161
    quartiles, 161
    range, 159
    relative standing, 161
    sample, *see* sample
    spread, 153, 159, 161
    standard deviation, 160
    variance, 159
studying, 2, 49
    time, 50

teams
    design, 186, 188, 191–193
    engineering, 5, 31, 69
    project, 16
teamwork, 31, 69, 184
technician, 2, 5, 6, 11
technologist, 2, 5, 6, 11
tenses, 71, 72
    chart, 72
thesaurus, 69–71
trade secrets, 204, 217
trademark, 35, 182, 204, 207, 214

certification, 215
distinguishing guise, 215
foreign, 215
ordinary, 215
truncation, 131
uncertainty
    absolute, 130, 132, 143
    derived, 114
    measurement, 114, 115, 125, 126, 129, 130, 147
    range, 126
    relative, 130, 133, 135, 143, 144
    written value, 129
Union for the Protection of Industrial Property, 215
unit systems
    absolute, 116, 120
    CGS, 116
    FPS, 115–117, 119–121, 123
    gravitational, 116, 117, 120, 121
    hybrid, 117
    SI, 114–116, 119–121
United Engineering Foundation, 47
United Kingdom, 21
United States, 7, 16, 17, 21, 42, 43, 45, 47, 60, 116, 120, 121, 123, 209, 212
units, 114, 115, 119
    algebra, 121
    Celsius, 121
    conversion, 121
    dimensionless, 121
    energy, 120
    Fahrenheit, 121
    force, 120
    fundamental, 115, 116
    imperial, 121
    kelvin, 121
    length, 120
    mass, 120
    power, 121
    pressure, 120
    Rankine, 121
    systems, 116
    temperature, 121
    time, 120
    U.S., 121
    work, 120
    writing, 118
university
    academic standards, 49
    applying, 7, 10, 20
    life, 49
    professors, 7, 13, 49–52, 54, 55
    student failures, 50
    studying, *see* studying

variance, 159
verb, 75
voice, 72, 73
    active, 72, 97
    passive, 71, 72
    tone, 76
von Oech, R., 196, 198

welder, 6
welfare
    individual, 45
    of engineers, 45
    public, 13, 19, 29, 32, 33, 70
whistle-blowing, 32
working environment, 16, 93, 218
writing
    aids, 93
    basics, 58, 69, 75
    brevity, 71
    clarity, 71
    English, 91
    errors, 69, 71, 75
    hints, 70
    instruments, 54
    mathematics, 77, 82
    steps, 93
    style, 58, 69
    technical, 58, 59, 69, 71, 73, 75, 77
    terms, 71
    test, 69, 70